THE
GEEK
GUIDE
TO
LIFE

ABOUT THE AUTHORS

COLIN STUART is a fellow of the Royal Astronomical Society and a physics and space geek. He has written for the *Guardian*, *New Scientist*, *BBC Focus* and the European Space Agency.

MUN KEAT LOOI is a science writer and features editor. He is the commissioning editor of *Mosaic*, the Wellcome Trust's unique publisher of in-depth science stories. He won silver Rising Star in the British Media Awards 2015.

 Twitter: @geekguidetolife

 Facebook: Geek Guide to Life

 Instagram: geekguidetolife

 Youtube: The Geek Guide to Life

This edition published in 2017 by SevenOaks
An imprint of the Carlton Publishing Group
20 Mortimer Street
London W1T 3JW

First published in 2016 by André Deutsch Ltd

Text and design copyright © André Deutsch Ltd 2016

Editorial Director: Piers Murray-Hill
Editorial: Anna Marx
Design: James Pople
Illustrations: Phil O'Farrell
Production: Sarah Kramer

All rights reserved.

A CIP catalogue for this book is available from the British Library.

ISBN 978-1-78177-709-1

10 9 8 7 6 5 4 3 2 1

Printed in Dubai

THE GEEK GUIDE TO LIFE

SCIENCE'S SOLUTIONS TO LIFE'S LITTLE PROBLEMS

COLIN STUART
MUN KEAT LOOI

SEVENOAKS

C O N T

1 HEALTH AND BODY

2 WORK AND CAREER

ENTS

C O N T

E N T S

INTRODUCTION

....

"BLESSED ARE THE GEEK, FOR THEY SHALL INHERIT THE EARTH"

GEEK *n*   1. a person who has excessive enthusiasm for and some expertise about a specialized subject or activities

Brawn may have won out in days gone by, but in the modern world it's brain that rules the roost. Even if those brains come with a penchant for science fiction, comic books and cosplay. Because it is the geek that has lovingly crafted our modern technological society. The Wright Brothers – geeks. Steve Jobs and Bill Gates – geeks to the core. Tim Berners-Lee, Mark Zuckerberg and Elon Musk – hardcore geeks of the highest order.

Geek was once a derogatory term – one thrown around to intimidate or ridicule those who didn't fit in. But today geek is chic. The word has been reclaimed as a badge of honour. Glasses are a fashion accessory, comic book heroes are big box office, scientists are the main protagonists in a smash hit comedy show. The video game industry is worth more globally than the film industry.

The passion, endeavour and intelligence of generations of geeks has led to a world of fast Internet, smartphones, tablets, driverless cars and virtual reality headsets. People have walked on the Moon and a man-made object has departed our solar system. Cancer rates are falling, smallpox has been eradicated and we're getting on top of HIV. Our unquenchable curiosity has led us to other planets, deep inside atoms, to the bottom of the ocean and far inside our DNA.

But scientists aren't just tackling these deep existential questions. They are also applying the scientific method to problems a little closer to home. And what they've discovered can help you navigate your way through life. Being alive today is a tricky business, particularly when there isn't an instruction manual for being a human. Until now that is. This is your manual. We've scoured the scientific literature so you don't have to. Pouring through journal articles and crunching through calculations, we bring you life lessons anchored in physics, chemistry, biology, psychology, economics and behavioural science. There is no self-improvement drivel here, just peer-reviewed or maths-backed research on the tips and tricks you need to get ahead.

Join us on a journey as we turn up the geek dial to eleven and guide you through the tips and tricks necessary to bolster the key areas of your day-to-day activities. Whether it's kitchen, commute, career or confidence, we've got you covered. Welcome!

CHAPTER 1

HEALTH
AND BODY

■ ■ ■

HOW TO COOL DOWN ON A HOT DAY – DRINK HOT TEA

When you're hot you reach for the cold drinks, right? Yet stories abound of people in humid countries like India reaching for a hot cup of tea rather than a cool glass of pop. What gives?

Scientists at the University of Ottawa in Canada put this weird anomaly to the test. Olly Jay and some colleagues got twelve men to cycle at a reasonable pace for seventy-five minutes. They were given three drinks – one cold (1.5°C/34.7°F), one at body temperature (37°C/98.6°F) and one hot (50°C/122°F). Everybody took one sip before starting and then at three points during the exercise. Jay and the team measured the cyclists' temperatures in different places around their bodies (if you must know, their skin, ear canals and rectums), as well as how much sweat they accumulated on their foreheads, upper backs and forearms (*see graph*).

Although body temperature stayed roughly the same regardless of the type of drink, the drinks did make a big difference to sweat. And this is the thing: those droplets dripping from your forehead and soaking up your shirt carry with them heat as they evaporate away from the body and to the outside. As much as it might feel icky, sweating is the best way for you to avoid meltdown. It's the body's most efficient method of cooling down.

And that's where the hot drinks come in. They obviously add heat, and this raises the temperature of your body, causing you to sweat more. And the loss of

heat caused by the sweating more than compensates for a little added warmth from the drink.

In a second experiment, the cyclists were given just the cold and hot drinks, but instead of swallowing them they were asked just to swish the drink around in their mouths for fifteen seconds. Others had the drink delivered directly to their stomach through a (nasty sounding) nasogastric tube. The point? To demonstrate once and for all that we have

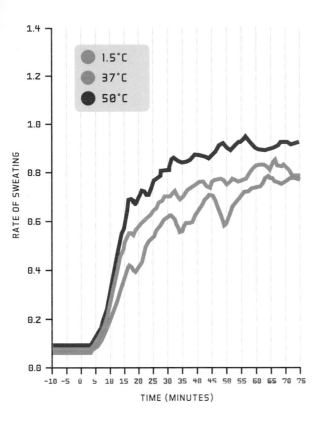

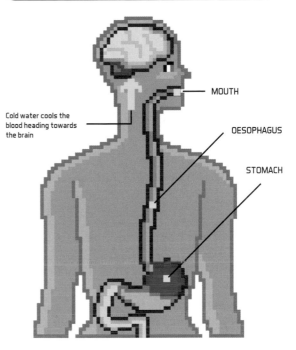

ABOVE: *Cyclist's rate of sweating vs time whilst drinking one of three drinks of different temperatures: cold (blue), body temperature (yellow), and hot (red).*

ABOVE: *There are internal body temperature sensors in the mouth, oesophagus and stomach, which are triggered by hot and cold drinks. Cold water cools the blood heading towards the brain, hence its perception of core body temperature.*

temperature sensors all over our body — both inside and out.

The swishing made no difference whatsoever to each cyclist's sweat rate. Only drinking it (that is, sending it directly to the stomach) did, indicating that the temperature sensors triggering the sweating mechanism are internal, within the stomach.

But wait! Before you go reaching for a thermos in the summer, it's worth remembering that while sweating cools you down, it also dehydrates you (since it's removing water from your body). So, some hot drinks, especially things like coffee, are actually a bad idea when your body is already producing a lot of heat, such as when you are cycling or doing some other exercise. In those circumstances,

you're likely to already be maxing out your sweat levels and just dripping it to the ground — which is much less effective than evaporation for cooling you down. And the effect on sweat rate is likely to change with different beverages. Something like a slushy (or indeed, a hot apple pie) causes much greater changes in body temperature. Studies have shown ice cream to cause a heat deficit five times greater than a cold drink, with a hot pudding loading on seven times more heat than a warm drink.

So, if you're jogging in the sun, an ice-cold bottle of water is still your best bet. But if you're leisurely sitting under a tree in the middle of summer, a hot tea can actually cool you down quicker.

HOW TO EAT LESS – BY IMAGINING EATING MORE!

We've all succumbed to the temptation of food at some time or another, especially when social norms tell us it's the place to do it. Take the cinema, for instance. It's tough to avoid giving into that box of popcorn when you're sitting in there, or that bag of chips you see on the way home. Resisting is a matter of willpower, but science tells us it's actually possible to avoid wolfing down the whole bag - using just a bit of imagination.

Experimental psychologist Carey Morewedge and colleagues at Carnegie Mellon University in Pittsburgh, Pennsylvania, asked fifty-one people to eat M&Ms and cheese (not at the same time!). They asked half the group to imagine eating thirty M&Ms. The other half of the group imagined eating just three M&Ms. Everyone was then allowed to eat as many M&Ms as they liked.

What did this achieve? Well, the people who'd imagined eating more sweets ate around three M&Ms, while the others ate five.

Having said that, before you start conjuring up the biggest bowl of chocolate-covered peanuts ever seen, it does depend on the type of food you're about to eat. The researchers repeated the same experiment on the M&M eaters, but this time gave them cheese. It made no difference – those who imagined eating thirty M&Ms ate just as much cheese as those who imagined three M&Ms. But when a group was told to imagine eating thirty cubes of cheese first, they ate much less of the real thing.

When asked, the participants said that the mental images didn't change how much they liked that food. However, it did seem to change how much they wanted it: in one last experiment the participants were asked to play a computer game to earn cheese

WHY SOME FILMS MAY MAKE US SNACK MORE

If you're looking to cut down on snacking at the movies, be sure to take note of what film you're watching. Research from Cornell University shows that audiences are more likely to munch on more popcorn while watching tearjerkers than they are while watching comedies.

Thirty people were randomly assigned to watch either the rom-com *Sweet Home Alabama* or the epic weep-fest *Love Story*, before switching to the other film the next day. At each screening they were given a bucket of popcorn and either a soft drink or water.

Surprisingly, viewers ate twenty-eight per cent more popcorn while watching the tragedy. Why? Well, possibly because popcorn is a comfort food. Physical or emotional distress is known to increase our intake of foods that are high in fat and sugar – such as popcorn. Foods filled with unhealthy goodness produce a feedback effect that inhibits activity in the parts of the brain that make and process the hormones that cause stress and related emotions.

cubes – and seemed to put in significantly less effort to get them.

The thinking behind these findings is that repeated exposure – in this case, to a particular food eaten bite by bite – reduces your desire for it. Psychologists call this habituation, and it seems to dampen the appetite separately from other physiological signals, such as blood sugar level or a bloated stomach.

Admittedly, daydreaming isn't going to have you losing vast amounts of weight anytime soon. But it's a useful exercise to cut down on snacking or indeed the amount you eat during meals. A few minutes imagining exactly what it is you're about to eat could counter your cravings, or at least mean you will be sated after just a bite or two.

HOW TO QUIT BITING YOUR NAILS

Though considered disgusting by some people, the habit of biting your nails is very, very common. Studies have suggested that as many as twenty per cent of Americans alone may do it - and it's a habit that human beings have indulged in since ancient Greek times.

It's harmless of course, although we didn't always think that way. Psychologists once considered nail biting, like hair plucking, to be a mild form of self-harm and a sign of self-loathing. However, more recent theories make a lot more sense – it's actually a distraction, a bit of stimulation or relaxation. After all, as researchers have observed, most people tend to do it when they're either under-stimulated (bored) or over-stimulated (excited or stressed) – and, of course, the hands are just *there*. All the time. So doing something with them provides either stimulation or relief, depending on the circumstance. The same line of thinking goes for smokers.

We're not the only ones in the animal kingdom to do such things, of course – think how much cats like grooming themselves by licking their fur. Studies have also found that many people with "body-focused repetitive disorders" such as nail biting tend to be classified as perfectionists – and, after all, trimming your nails can be immensely satisfying. Do it a lot, however, and such behaviour can reinforce itself over time.

So we think we know why we do it, but is it actually *wrong*? Generally no, but your hands, being what they are, of course pick up all kinds of germs from all kinds of places that tend to accumulate beneath the nails, and biting them can give those germs a free ride straight to your gut. Tears in the skin could also lead to infections. So, if you suffer from this habit and you think it is becoming extreme, you should probably try to stop.

Psychologists have suggested a lot of different techniques to help you give up nail biting, but it all goes back to the source. First, think about the trigger – when is it exactly that you end up biting your nails? Now try to change the circumstance, or seek out an alternative. Love to bite your nails while watching television? Chew some gum, or play with a ball, instead. Try putting up little notes telling you to stop, as well – they can serve as useful reminders of your desire to stop biting.

If it's an emotion or a specific feeling – such as frustration – that's the cause of your nail biting, recognize when you feel like that and then consciously give yourself something else to do: clean the kitchen, or go for a walk with your hands in the deep pockets of a sweatshirt or hoody. Alternatively, you could try to make it physically harder to bite your nails – some sufferers have even deliberately applied clear nail polish that tastes awful.

See what works for you – a combination of some or all of these techniques may just do the trick. The important thing is to keep at it: just because you give in once doesn't mean you've failed. The longer you can refrain from biting each time, the surer you are to break the habit.

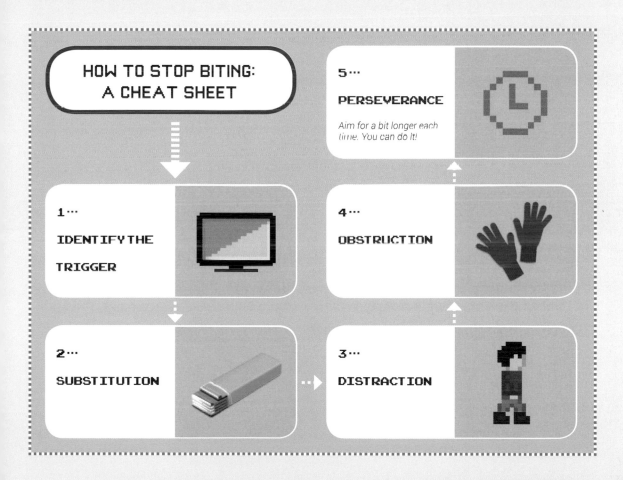

HOW TO STOP BITING:
A CHEAT SHEET

5···
PERSEVERANCE
Aim for a bit longer each time. You can do it!

1···
IDENTIFY THE TRIGGER

4···
OBSTRUCTION

2···
SUBSTITUTION

3···
DISTRACTION

HOW TO CURE
A HANGOVER

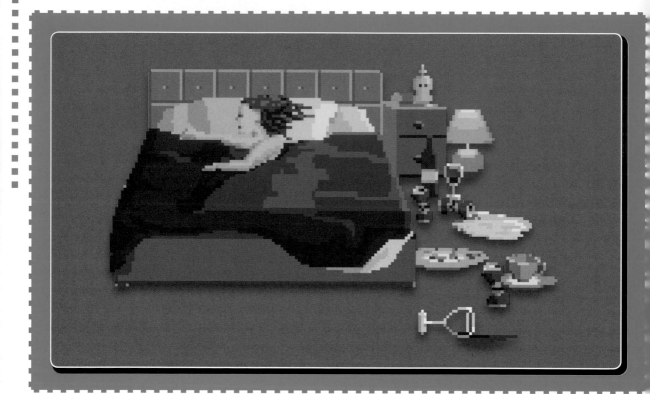

If you're old enough to drink alcohol, chances are you're old enough to have experienced the agony of a hangover. Why, after you've long stopped inebriating your body, do you still feel a bucket-load of awful symptoms - from headaches to dizziness, tiredness, stomach pains, sweating, thirst and nausea?

Sadly, science doesn't fully understand why. But it's come up with some pretty good ideas so far. First, alcohol is, of course, a poison. And the first by-product of breaking down alcohol is acetaldehyde, which is ten to thirty times more toxic than alcohol itself. Studies have found that people with hangovers not only have higher levels of acetaldehyde in their blood, but high levels of cytokines – signalling molecules used by the immune system in responses such as inflammation that trigger fever, usually to fight infections. If excessive drinking really does lead to a flood of cytokines, this could account for your muscle aches, fatigue, headaches and nausea, amongst other things.

The severity of a hangover also depends on the person having it. Body size makes a difference to how much alcohol you can cope with (that is, break down). And if your background is east Asian, unfortunately you probably have a mutation in the gene that makes alcohol dehydrogenase – the enzyme that breaks down alcohol – giving you a turbo-charged version that's super effective at making acetaldehyde. Even more unfortunately, some east Asians also have a mutation that slows the next step – turning acetaldehyde into acetic acid – so that they have to put up with an excess of the toxin for longer, resulting in the common flush of many drinking Asians, as well as a terrible hangover.

But, I hear you cry, how do I get rid of my hangover? Sadly, science doesn't have a clear answer to that question. The most effective solutions are also the most obvious: avoid alcohol in the first place, or drink moderately and while pacing yourself, especially if you're of a smaller build. Drink lots of water – alcohol is a diuretic (that means it increases the production of urine), and the chances are that if you've been downing beers or wine all night you won't have drunk enough water to counter this. That's what causes the terrible thirst and light-headedness that come with a hangover.

The old adage of never drinking on an empty stomach also holds up, though not for the reason you'd think. The food doesn't absorb the alcohol, but filling your digestive system slows the rate at which your body will absorb the drink.

There's also a good case for watching what you drink. Some drinks, such as whisky, obviously contain more alcohol than beer, especially if you're drinking it neat (undiluted), without a mixer. Some drinks also have higher levels of what are known as "congeners" – trace chemicals that are produced during fermentation. Studies have shown that darker-coloured liquors, such as bourbon, which have high congener levels lead to more severe hangovers than light-coloured or clear liquors like vodka, which have none.

One particular congener called methanol – which is found in the highest quantities in whisky and red wine – can linger in the body after all alcohol has been eliminated, perhaps accounting for the enduring effects of a hangover. This could also explain why many people say mixing drinks is a really bad idea; a greater variety of congeners could well lead to a wider variety of effects. It doesn't explain any supposed benefits of drinking in a particular order though, so the old saying "beer before wine is fine" is actually a myth.

7 SCIENCE-BACKED WAYS TO AVOID A HANGOVER:

1. Don't drink.
2. If you must, only drink moderately and pace yourself.
3. Remember who you are, especially your body size – if you're small, don't try to match a larger person drink for drink.
4. Watch what you drink – and don't mix.
5. Make sure you eat.
6. Drink plenty of water.
7. Rest.

THE GEEK GUIDE TO COLDS AND FEVERS

WILL I CATCH A COLD FROM WET HAIR?

Old wisdom warns that you'll catch your death if you go out in the cold with wet hair. But any good geek knows that a "cold" is actually caused by a virus. So, does rushing outside straight after taking a shower really increase the risk of you catching a cold?

The short answer is no, not really. In the name of science, scientists have deliberately lowered people's temperatures and then exposed them to the cold virus (all participants were volunteers, we might add!). However, such studies haven't come up with anything conclusive.

On the other hand, other experts have looked at whether getting cold and damp activates the virus. Researchers at the Common Cold Centre (yes, that really exists) in Cardiff, UK, had 90 out of 180 people (again, all volunteers) soak their feet in cold water for twenty minutes, while everyone else kept their shoes and socks on and just sat in an empty tub. When the

researchers checked on everyone four or five days later, twice as many in the chilly footbath group said they had caught a cold.

It's worth saying that this was only what the participants reported – no medical tests were done to confirm that they were definitely infected with the virus. However, anecdotally at least, being cold and wet seems to have a significant effect on whether or not you catch a cold.

The theory goes that when your body temperature drops, blood vessels in your nose and throat tighten. As well as carrying blood, these passageways are also the principal thoroughfares for the white blood cells – which are quite literally your bodyguards against infection. If fewer of them are getting to the frontlines because the roads are closed, then this could lower your defences. When you're warm and dry again, everything opens up and the white blood cells can get to where they're needed. But by then, the virus may already have broken in and started multiplying – hence the sneezing, blocked or runny nose and cough as your body gears up to fight and eject the invaders.

So, as far as the evidence goes, you won't "catch" a cold from wet hair as such, but it is possible that it might activate a dormant virus that is already nestling in your body. No matter how much of a rush you're in, or how lazy you're feeling, it's probably best to dry yourself properly and wrap up warm before you head outdoors.

SHOULD I BLOW MY NOSE?

Maybe this sounds like a stupid question. You've got a cold, you're bunged up, you can hardly breathe and you feel like crap. Why shouldn't you blow your nose?

Scientists at the University of Virginia wanted to find out. They did what anyone would: hooked up four volunteers to measure the pressure in their noses as they sneezed, coughed and, yes, cleared their noses.

Coughing and sneezing causes as much pressure as just breathing, but nowhere near as much as blowing your nose. It's enough to project one millilitre of mucus not just out of your nose but actually backward into your sinuses – every time you blow.

How much of a difference this makes scientists are not sure, but the researchers did suggest that it might indeed counteract your body's natural mucus drainage system. They were more concerned about whether it could introduce viruses or bacteria into the inner reaches of your nose.

Of course, we can't help wanting to blow our noses. And since mucus is the body's way of ejecting viruses and bacteria, it does make sense. So if you have to blow, blow gently, and one nostril at a time. Alternatively, sniff and swallow.

SHOULD YOU REALLY "FEED A COLD AND STARVE A FEVER"?

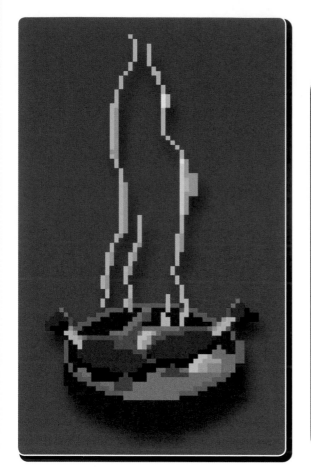

	COLD	FEVER
EAT PLENTY	✓	✓
DRINK LOTS OF WATER	✓	✓
REST	✓	✓
EAT SOUP	✓	✓
DRINK HOT LIQUIDS	✓	✓

It's one of the oldest of old wives' tales – if you're ill with a cold, eat to generate heat, but if you're feverish, starve to drop your temperature. On the face of it, this sounds entirely logical, but here comes science to show that this theory is, in reality, completely untrue.

The fact is, no matter how sick you are, you should eat. Your body needs energy to fight any illness, so eating well is an absolute must. Yes, eating does keep you warm, generating energy and body heat as you digest, while also converting stored energy to fat. Thus,

there is some truth to the old proverb when it comes to colds. However, fevers are a slightly different matter.

A fever is actually part of your body's natural reaction to illness. The scientific reasoning behind this is that raising body temperature helps you kill off bugs. However, doing so also increases your metabolism and burns calories, with the energy demand escalating with each degree of an increase in temperature. Consequently, it's extra important to eat to help meet this need.

Drinking is even more vital – you probably recall how sweaty you become when you are either feverish or suffering a cold, so replacing these fluids is crucial in either case, even if you may not feel like eating or drinking.

Taking in water can also help get rid of the mucus that causes your coughing and sniffing. The production of mucus is the body's natural way of trapping and expunging germs. However, when your lungs, nose and throat are dry, mucus can block things up – especially when it hardens. So, stay hydrated to keep that icky mucus flowing – it's a necessary evil.

A good solution to all this might be to eat soup. What we've learned probably accounts for why another old wives' tale – that eating chicken soup cures all ills – holds up as a traditional remedy for colds and flu. It's high on calories, replenishes your liquids, and the warm vapour is useful for, ahem, loosening your dried mucus. In fact, the same goes for any kind of soup or hot drink, so why not take the opportunity to enjoy a nice warm brew?

WHY ILLNESS TENDS TO RUIN YOUR HOLIDAYS

We've all been there. After a long, hard innings at work, you've finally persuaded the boss to give you two weeks off. You skip out of the door, luggage in tow, running gleefully to the airport... and then you sneeze. By the time you've landed you're feeling lethargic and irritable. The immigration officer keeps her distance from the snot running down your nose. You spend the next few days confined to your hotel room, tissues stuffed up your nose and lacking the energy to do anything other than watch daytime TV.

This is what psychologists have labelled "leisure sickness" and it strikes with uncanny timing to ruin our precious annual leave. Science has its theories as to why. One line of thought is that you simply let your guard down. Of course, your immune system is running like crazy, particularly when the prospect of a vacation is accompanied by a frantic rush to tie up all loose ends before you leave – not to mention pack and prepare. But there's never actually been any scientifically proven evidence for this.

More likely, itchy feet, mass air travel and globalization have made travellers of us all – and the bugs are often hitchhiking a ride. Think about it. You probably spend the first part of your holiday in a teeming hotspot of new bugs brought from all over the globe: the airport; in lines; at security; at the shops; in the toilets.

After a few hours seasoning there, you finally make it onto the plane – holed up in a sealed metal tube, possibly for even longer than you'd been at the airport, cooked up with several hundred other passengers in various degrees of health. And in much closer proximity, for longer periods of time than you'd normally prefer. Up in the air, the temperature drops, creating the perfect environment for the common cold and influenza that, for yet unknown reasons, seem to thrive in cool conditions. Air conditioning helps to filter, but it also tends to dry things out. And low humidity helps promote viral transmission. Plus, there's not much you can do about the people in your immediate vicinity. And in airspace, there's nowhere to run. You're stuck there with whatever your fellow passengers are coughing out, not to mention spreading to the seats, doors, curtains and other fixed objects around you.

Science has long known how increased global travel has been a boon for viruses as much as travellers. Unfortunately, there's not a lot you can do about this. If you want to travel, you'll just have to get your flu shot, vaccinations and take your chances.

WHY YOU SHOULD AVOID BEING "HANGRY"

You know the feeling: to be "hangry" is to be irritable, largely because your body is craving nutrition. Yet "hangriness" isn't just a new fad thought up by hipsters and new-age medics: science shows that being hangry is a real thing, and the swing in your mood can have consequences - not least for the person that has to see you almost every day.

In a 2014 study, 107 couples agreed to be studied for three weeks, taking part in various activities to judge their moods. The researchers, led by Brad Bushman at Ohio State University, found that partners would get increasingly angry and mean towards one another when their blood sugar was low (if you are wondering how they measured meanness, they gave the subjects voodoo dolls of their significant others in which to stick needles. Hangry spouses were more likely to stick the pins in). They were also more aggressive – measured by seeing how long a subject would force their spouse to listen to loud, unpleasant sounds, such as fingernails on a chalkboard and dentist drills.

In a separate study, also led by Bushman, sixty-two volunteers were given a drink and then competed anonymously against a same-sex partner in a game. Crucially, the participants could set the decibel level of a blast of white noise their opponent would receive as a forfeit each time they lost – a measure of aggression, in the laboratory, anyway. Those who were given lemonade set a kinder forfeit than those who drank a placebo.

In related studies, subjects given sugary drinks were less likely to seem frustrated when playing an impossibly difficult game. Other researchers have

1

GRRR...

LOW BLOOD SUGAR AND IRRITABLE – THE RED MIST DESCENDS...

tried deliberately reducing people's blood sugar by giving them insulin – leading to subjects reporting more negative moods.

Why would this be? Well, the brain is the most energy-hungry of your organs, and simple sugars such as glucose are what fuel it. When the amount of sugar in your bloodstream goes down, so does your brain's ability to function – a problem when you are trying to concentrate, practise self-control or do other brain-intensive work. Data from alternative studies suggest that those worse at regulating blood sugar are more likely to exhibit bad behaviour. However, let's not forget that correlation does not always mean causation. Different people are better or worse at regulating glucose levels, so low blood sugar affects some more than others. And this can be influenced by many different factors, such as your genes. Nevertheless, awareness of the link means you can do something about it. All the more reason to reach for a small snack when you feel the red mist descending...

REACH FOR A
SNACK BEFORE
THIS TURNS NASTY

HAPPINESS RESUMES
AT GEEK TOWERS

HOW MUCH EXERCISE YOU REALLY NEED

Sure, exercise is good for you: it improves your mood, boosts your immune system and keeps your weight down. Moreover, the benefits are solid: one study of 650,000 people in Sweden and the United States found that those who did a moderate amount of exercise (thirty minutes a day, five times a week), lived on average 3.5-4.5 years longer than those who did none.

It may be good for us but many of us just don't feel like an hour in the gym, even if we know it's important. The good news is, you could get by on much less than that and still reap the rewards.

Chi Pang Wen and a team at Taiwan's National Health Research Institutes wanted to see just how little. They studied 400,000 people over eight years and found that those who did just fifteen minutes of physical activity per day had a life expectancy around three years longer than those who did nothing. The results shone through even when differing lifestyle factors such as smoking, drinking and diabetes were taken into account.

The stats showed they were fourteen per cent less likely to die in the study period than those who did nothing, and that one in every nine deaths from cancer in the inactive group could have been avoided. Moreover, each additional fifteen minutes a day of exercise brought added benefits – though it was the first fifteen minutes that seemed to make the biggest difference.

The better news is that it is general activity that makes the difference, as opposed to concentrated exercise. This means that your daily exercise doesn't have to be that run in the cold you've been dreading or a major slog through a sweat-fest at the gym.

TOP TIPS:

1. **Anything physical counts, so walk the dog, go dancing or clean the house – it doesn't have to be the gym.**
2. **Do exercise in moderation.**
3. **Don't forget to rest – it's just as important as exercise.**

Experts suggest that a brisk walk – the kind of pace you might use if you were late for a date – is enough. Or take the dog out to the park, go dancing, play with your kids – the key is finding something you like to do, so that you will do it regularly.

Of course, there are those who like harder workouts, and those people achieve the same health benefits in half the amount of time. People get a buzz out of working out and that's generally a good thing, although some studies have suggested that too much exercise can be harmful. More strenuous jogging may, for instance, put a strain on the heart to an extent that cancels out the benefits of exercise. The jury's still out on whether that's really true, but what is known is that there are physiological reasons why rest is just as important as exercise.

First, you need to rest in order to get stronger. "Rest" is actually the state in which your body repairs muscle fibres that have been damaged during exercise, stitching them together to make stronger muscles. Round-the-clock training doesn't allow for this and can lead to injuries such as tennis elbow or tendinitis. It can also lead to fatigue, making it more likely you'll pull a muscle, fall over or cause some other type of injury, not to mention exhausting yourself mentally. Exercising more, by its nature, offers more opportunity to suffer injury, particularly if you're tired.

So know your limits, but also pace yourself. That means building up your workout gradually, too – don't go from 0 to 100 right from the get-go. Ultimately, what counts is not how much exercise you are actually doing, but that you are doing it in moderation.

JUST FIFTEEN MINUTES
OF EXERCISE A DAY COULD BE ENOUGH TO INCREASE YOUR LIFE EXPECTANCY BY THREE YEARS.

HOW MUCH SLEEP YOU REALLY NEED

We all love to sleep, but many of us struggle both with getting to bed early and getting out of it in the mornings. Throw in coffee, energy drinks and artificial lights - including smartphones - and it's no wonder that getting the recommended eight hours of rest is a rarity for most of us, despite the fact that we spend on average up to a third of our lives asleep.

The consequences can be considerable: loss of concentration, depression and anxiety – even a risk of diabetes, high blood pressure and obesity, if the lack of sleep continues. Yet too much sleep is associated with the same problems. So, as a narcoleptic Goldilocks might ask, how much sleep is just the right amount?

The US National Sleep Foundation waded through 320 research papers to come up with the most up-to-date guidelines for how much you might need, depending on your gender, age and lifestyle.

For adults – and by that we mean people aged eighteen to sixty-four – between seven and nine hours each night will do you fine. Of course, how much sleep

RECOMMENDED HOURS OF SLEEP

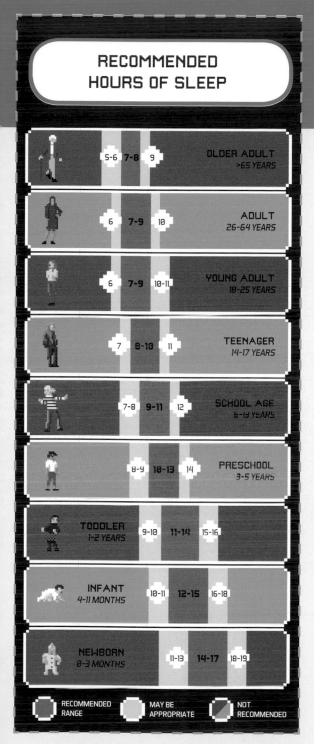

OLDER ADULT >65 YEARS	5-6	**7-8**	9
ADULT 26-64 YEARS	6	**7-9**	10
YOUNG ADULT 18-25 YEARS	6	**7-9**	10-11
TEENAGER 14-17 YEARS	7	**8-10**	11
SCHOOL AGE 6-13 YEARS	7-8	**9-11**	12
PRESCHOOL 3-5 YEARS	8-9	**10-13**	14
TODDLER 1-2 YEARS	9-10	**11-14**	15-16
INFANT 4-11 MONTHS	10-11	**12-15**	16-18
NEWBORN 0-3 MONTHS	11-13	**14-17**	18-19

RECOMMENDED RANGE   MAY BE APPROPRIATE   NOT RECOMMENDED

one needs to feel fully rested depends very much on the individual. For those over sixty-five years of age, the recommendations are adjusted slightly to seven or eight hours. Kids (six to nine years old) generally need nine to eleven hours. For teenagers, eight to ten hours a night will normally suffice.

When it comes to sleeping, the most important thing is to prioritize your beauty sleep. It's not what you do after everything else is done – it's the rest you need to do anything properly.

SEVEN TIPS FOR BETTER ZZZZZ...

1. Stick to a sleep schedule, even at weekends.
2. Practise a relaxing bedtime ritual, e.g. reading a book.
3. Exercise daily.
4. Evaluate your bedroom: get temperature, sound and light right for you. Sleep on a comfortable mattress and pillows.
5. Beware of hidden sleep stealers, such as alcohol and caffeine.
6. Turn off electronics before bed – don't use them before sleep and make sure they don't disturb you during it.
7. Prioritize sleep!

WHY YOU SHOULDN'T LIE-IN AT THE WEEKEND

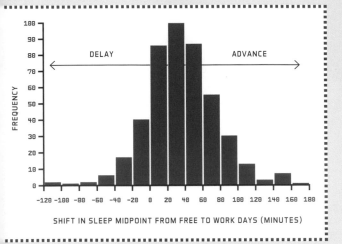

DELAY ← → ADVANCE

FREQUENCY

SHIFT IN SLEEP MIDPOINT FROM FREE TO WORK DAYS (MINUTES)

LEFT: *The difference in sleep between weekdays and weekends shifted the participants' sleep cycle (measured by plotting the "mid-point" of their sleep) by two hours or more.*

It's a fair question for any hard-working geek: after a week waking at the crack of dawn, don't I need (rather, deserve?) a lie-in to catch up on my beauty sleep? However, if you're in this routine, sadly, you might be doing yourself more damage than you think.

When you're switching sleep schedules – say from 11pm–6am to 2am–10am – from weekday to weekend and back again, you're effectively moving across three time zones. Yes, you're effectively giving yourself jetlag!

Psychologists at the University of Pittsburgh in the US studied 490 people using sleep monitors (worn on their wrists), blood tests and health assessments. Most tended to stay up a bit later and sleep in a bit longer at the weekend. No big surprise there. But those who had the biggest shifts in their sleeping pattern also tended to have larger waists, higher BMI (body mass index) and signs of things such as insulin resistance that are warning indicators for a higher risk of diabetes, obesity and heart disease. The link stayed strong, even after accounting for differences in smoking, exercise and other behaviours. Meanwhile, another European study of 65,000 people found that a third of those suffering from two-plus hours of sleep misalignment were also more likely to be classified as overweight. Studies have

also found that lack of sleep disrupts the chemical balance of our brains, triggering the munchies. The delightfully named substance Endocannabinoid 2-AG makes it pleasurable to gorge on high-fat foods, and a sleep-deprived brain finds itself swamped in more of it than usual.

By sleeping well our body is able to keep its own internal body clock and prepare itself for the day ahead – producing more insulin in the morning, for instance, in preparation for breaking down breakfast. This sense of time (known as the circadian rhythm) regulates our organs, systems – even the expression of our genes. Animal studies have shown that disrupting this vital rhythm can lead to toxins building up in cells. By disrupting that pattern, we are eating when our bodies aren't ready, and sleeping when we're expecting to be awake. It's a burden that takes its toll.

Don't rely too much on coffee when you're sleepy, either. Studies have calculated that the equivalent of a double espresso just three hours before sleeping can set your internal body clock back by around forty minutes.

HOW TO TAKE THE PERFECT NAP

Bill Gates swears by a power nap, and its benefits are proven by actual research. Napping can do you wonders, from increasing your alertness to improving your memory. So let's do it right:

1. Doze in the afternoon. Your energy levels tend to dip around six to seven hours after waking – about 2pm for most people. This ideal "hour of the nap" is also far enough away from your bedtime not to interfere with your sleep cycle.

2. Go somewhere dark, comfortable and quiet. Research has established that these obvious factors have a massive effect on how easily and quickly you fall asleep.

3. Calculate how long to nap. When sleeping, your brain goes through a series of distinct stages, so naps of different lengths can have very different effects: longer naps produce longer-lasting benefits, but you run the risk of intense grogginess if you are disturbed too early. The best naps tend to be those of thirty minutes or less; with these, your brain never advances past the early stages of light sleep and you get a burst of improved alertness. However, if you're watching the *Lord of the Rings* trilogy (extended editions) back-to-back, prepare for a longer nap that hits the deeper, higher-quality stages.

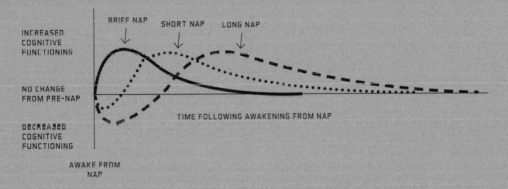

4. Set an alarm! This will help you avoid over- (or under-) sleeping.

5. Drink coffee. Seriously. Down an espresso (or other caffeinated beverage) and then take a twenty-minute snooze. Studies indicate that you will wake up more alert and perform better on memory tests than if you'd done either alone. It sounds nuts, but the mythical coffee nap has a real scientific basis. Caffeine makes you feel more energetic by displacing adenosine, the chemical that makes you feel tired. It takes around twenty minutes for caffeine to move through your digestive system to the bloodstream and enter the brain. Now, consider how sleep naturally clears adenosine from your brain: wake up after twenty minutes when the caffeine is arriving in your brain and *voila!* – caffeine turns up with no adenosine to even displace, just a clear run to the brain receptors, amplifying its effect.

CHAPTER 2

WORK AND CAREER

WHAT'S THE BEST WAY TO COMMUTE TO WORK?

For many of us, the commute is the worst part of the daily grind. You have to fight your way through traffic jams or packed trains before you even start work in earnest. Then, at the end of a long day, you have to do it all over again.

In the United States, the average travel time to work is twenty-five minutes – that amounts to a total of just over four hours a week spent going back and forth to work. Unsurprisingly, New York is the city with the longest average commute. Residents of the Big Apple spend six hours and eighteen minutes per week travelling. That is 12,600 hours over an average working lifetime – almost a year and a half of your life spent commuting. It isn't any better in Britain, where more than three million people spend over two hours a day on the move.

According to researchers at the University of Waterloo in Canada, these substantial commutes are having a negative effect on our well-being. The researchers reported that the longer someone's commute, the lower their level of life satisfaction. This appears to be backed up by a 2014 British survey by the Office of National Statistics. A 2011 study from Sweden also links lengthy commutes with decreased energy, increased stress and higher work absences due to illness.

Given how much of our lives we spend doing it, and that for many of us commuting is a necessary evil, is there an optimum way to commute in order to limit the damage? It probably won't come as much of a surprise that driving appears to be the worst option. In August 2015, a study published in the journal *Transportation Research* compared the stress levels of driving, taking public transport and walking to work. Of the four thousand participants, it was the walkers who reported the lowest levels of stress and the drivers the highest.

It also seems that commuting regularly by car is bad for your physical health. A 2012 study published in the *American Journal of Preventive Medicine* found that the longer you drive, the worse your blood pressure and Body Mass Index (BMI) become, even if you undertake physical activity in other ways. An earlier study published in the same journal found that every hour spent in a car makes you six per cent more likely to be obese.

If you want to avoid driving, what else should you be doing? A 2014 study by medics at the University of East Anglia, UK, examined eighteen years' worth of data encompassing information on eighteen thousand commuters who had changed their mode of transportation. They found that even switching from driving to public transport was better for overall well-being. Those swapping to walking or cycling felt under less strain and reported higher levels of concentration at work.

So, while long commutes are on the increase in this modern, fast-paced world, if you can avoid driving to work, you probably should. And if you must drive, keep it as brief as possible.

HOW JOURNEY TIME AFFECTS HAPPINESS

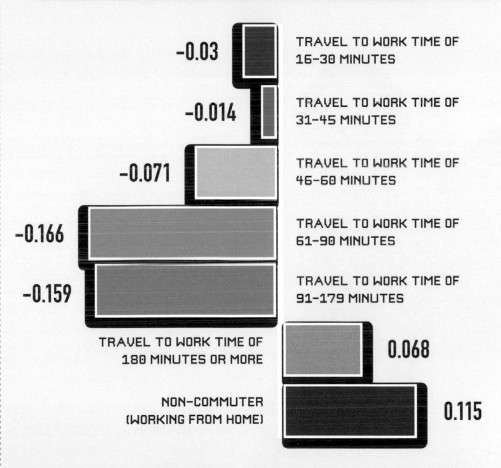

-0.03 — TRAVEL TO WORK TIME OF 16-30 MINUTES

-0.014 — TRAVEL TO WORK TIME OF 31-45 MINUTES

-0.071 — TRAVEL TO WORK TIME OF 46-60 MINUTES

-0.166 — TRAVEL TO WORK TIME OF 61-90 MINUTES

-0.159 — TRAVEL TO WORK TIME OF 91-179 MINUTES

TRAVEL TO WORK TIME OF 180 MINUTES OR MORE — 0.068

NON-COMMUTER (WORKING FROM HOME) — 0.115

This compares to a reference group of those who take one to fifteen minutes to travel to work.

HOW TO BE MORE CONFIDENT

At times, confidence can seem a bit like the Loch Ness monster or the Abominable Snowman - others have claimed to see it, but you're not always convinced it is really there...

However, even if you don't feel very confident, scientific studies have shown that there are ways to "trick" yourself into boosting your confidence. One of these is a particularly good ruse to try ahead of big presentations, an important meeting or the interview for your dream job.

Firstly, examine your body language. Standing up straight and tall not only makes you look more confident to others, but it also has the same effect on you. Social psychologist Amy Cuddy from Harvard Business School calls it a "power pose". In 2010 she published a paper in the journal *Psychological Science* that examined the effect of assuming a powerful stance – essentially standing like Superman or Wonder Woman for two minutes. According to Cuddy's findings, changing the shape of your body in this way can increase testosterone in your body as well as reducing cortisol (a stress hormone). She reported that it is not just your physiology

STRIKE A "POWER POSE":
STANDING IN A POWERFUL WAY MAKES YOU FEEL MORE POWERFUL.

that changes, either – power posing also makes you *feel* more powerful. Later studies have questioned the physiology claims, but the idea that standing in a powerful way leads to you feeling more powerful appears to be robust. So, next time you are preparing for a big meeting, duck into the toilets or stationery cupboard and stand like a superhero!

While you're at it, you should consider donning some headphones and pumping out a bit of music with a powerful bassline. That's according to a study by the Kellogg School of Management at Northwestern University. First, they played a group of participants a number of songs and asked them to rate how powerful, dominant and determined the songs made them feel on a seven-point scale. The songs that came out on top included Queen's "We Will Rock You", 2 Unlimited's "Get Ready for This" and 50 Cent's "In Da Club". These songs were then played to a new group of participants, with another new group hearing the songs rated lowest in the first task. These included Notorious BIG's "Big Poppa", Fatboy Slim's "Because We Can", and Baha Men's "Who Let the Dogs Out?"

The two groups of participants were then asked to fill in the blanks in the word P_ _E R with the first thing that came to mind. Those listening to the powerful songs were more likely to complete it as POWER compared to the other group. The group played the powerful music were also more likely to volunteer to go first in a debate. Intrigued by their findings, the researchers took the powerful songs and either made the bass more powerful or dialled it down. The researchers found that the bass-heavy versions came out on top.

If you're obsessed with a celebrity, thinking about them might also help you combat low self-esteem.

In 2008, a team led by psychology professor Shira Gabriel published research in the journal *Personal Relationships* encompassing three scientific studies looking at a total of 348 university undergraduate students. Participants were asked to write an essay extolling the virtues of their favourite famous person and also filled in a questionnaire designed to look into their self-esteem. Before putting pen to paper, those with low confidence were more likely to see their celebrity as close to their ideal selves rather than their actual selves. But after the writing task the researchers found that low-confidence participants reported feeling closer to their ideal selves and experienced a boost in self-esteem. High-confidence participants did not experience such a change.

So you want to make yourself feel more confident? Stand like Superman, think about your favourite celebrity and listen to Queen.

PITCH PERFECT

Your ability to feel more confident and powerful may be influenced not just by what you say but how you say it. In a 2012 study, published in the journal *Social Psychological and Personality Science*, participants were split into three groups based on how they were asked to read a passage of text. One group was asked to read it silently, with the other two reading it aloud either in a higher or a lower voice than normal. Those speaking in a lower pitch were subsequently more likely to choose powerful adjectives from a list than the other groups. The researchers concluded that perhaps speaking in a lower voice makes us feel more powerful, at least subconsciously.

HOW TO BE A BETTER PUBLIC SPEAKER

If you are afraid of speaking in public, then you certainly aren't alone. Survey after survey has shown it to be one of the things that people fear the most. In a now infamous poll from the 1970s, forty-one per cent of people said they were afraid of public speaking, whereas only nineteen per cent were afraid of death.

- -

Later surveys came to a similar conclusion – that just under half of us dread standing up and talking in front of people. One poll found that only snakes terrify us more.

What is it about public speaking that gets a lot of people so riled up? Psychologist Matthias Wieser and his colleagues believe they may have found the answer. In a 2009 study, published in the journal *Psychophysiology*, they suggest that the anxiety of being in the spotlight means we become more attuned to angry faces. Participants were split into two groups, one of which was told they would shortly be required to give a two-minute speech on a controversial subject that would be judged by a panel of experts. The second group was told that at some stage they would have to write an article about a non-controversial topic.

All participants were then hooked up to EEG machines to measure their brain activity and shown ninety-six flashing images of strangers – a mixture of happy, angry and neutral faces. Those in the public speaking group reacted more quickly to angry faces than their counterparts in the writing group. The authors concluded that we pick up on angry faces more swiftly when we are anxious, re-enforcing the fear of public speaking.

Being more sensitive to the angry members of an

audience is likely to mean we overlook the positive feedback being provided by others. If that makes us more fearful of public speaking, then we are likely to be even more anxious next time around. It's a vicious circle, but help is at hand. According to Daniela

Schiller's 2010 paper in the journal *Nature*, we can rewrite our fears, as long as we are exposed to them again soon after. So, swiftly get back on the horse by redoing the offending presentation, this time in front of a friendly audience who will nod and smile along.

Clearly, many of us find public speaking a challenge – and it seems that might be hard-wired into the way our brains work – and yet it is a key part of career success. So, what else can you do? First, ignore the old cliché that imagining the audience in their underwear will help your confidence as you prepare to speak. There is absolutely no scientific evidence to back up this assertion, and it is likely to be more of a distraction than a help. However, another form of visualization has been shown to be helpful in tasks like these. You just have to do it *before* your presentation, not during.

It is called "process visualization". The key is not to picture the outcome of your talk (it going well, people applauding or patting you on the back afterwards) but the steps required for that to happen. That is, you preparing and rehearsing the talk. Then you delivering it to an audience, some of whom will look bored and disinterested, while others lap it up. This approach is backed up by a study published in the journal *Personality and Social Psychology Bulletin*, which showed that students who visualized studying effectively got more A grades than those who visualized getting an A. Likewise, a study published in the *Journal of Sport Behavior* showed that tennis players who imagined the steps required to get better improved their game more effectively than those who simply imagined they were better at tennis. In short, to be a better public speaker, visualize the steps that you need in order to improve.

If all else fails, then you might consider having lots of sex in the run-up to a big presentation. But it has to be penetrative sex – other forms of intimacy don't quite cut it. At least that is according to psychologist Stuart Brody. He asked participants to keep sex diaries for a fortnight, logging occasions they indulged in penile-vaginal intercourse, sexual activity with a partner without such intercourse, or masturbation. They were then asked to perform a public speaking task. Brody's conclusions, published in the journal *Biological Psychology*, found that the participants who were least stressed out by the task were the ones who had exclusively had penetrative sex over the previous two weeks.

WHAT MAKES A GREAT TALK?

TED talks are some of the most watched public speaking content on the Internet. The most popular have been viewed tens of millions of times. That's a rich seam of information to mine for insights into what makes a great talk, and Vanessa Van Edwards did just that. She recruited 760 volunteers to watch videos posted on TED.com and quizzed them on what they saw. Here are some of her key findings:

• It is just as much about how you say something as what you are saying. The participants gave the same speaker near identical ratings for charisma, intelligence and credibility, even when the sound was muted.

Take home message: watch your body language.

• Your hands are key. Van Edwards found that the more hand gestures the speaker used, the more views the video had. Handsy speakers were also rated more highly on charisma.

Take home message: get your hands out of your pockets and use them to illustrate your points.

• Smile, even if you are talking about a serious subject. Those who smiled for at least fourteen seconds during their TED talk were rated as more intelligent by the participants.

HOW TO IMPROVE YOUR MEMORY

Have you ever been in an important meeting, only for the name of your most valuable client to slip out of your head? What about being at a conference and bumping into someone who clearly knows who you are but you can't for the life of you remember anything about them? If so, then you might need to brush up on your memory skills.

According to science, the key to remembering a person's name is to repeat it to someone else. That is according to a 2015 study by Victor Boucher, published in the journal *Consciousness and Cognition*. All participants wore headphones playing white noise whilst being shown words on a screen. They were split into four groups, depending on what they were asked to do next: say the word in their heads; silently repeat it whilst moving their lips; say it out loud; or say it out loud to someone else. They were later asked to pick out the words they had seen from a longer list. Those who had said the words out loud to another person were the most likely to remember them. So, if you want to get better at recalling someone's name, perhaps consider using it in a sentence when talking to them. "Yes, Jane, I can do that for you" has more chance of sticking than "Yes, I can do that for you." Alternatively, be name-specific when telling a partner or friend about your day.

That cup of coffee can help, too. A 2013 study in *Nature Neuroscience* led by Daniel Borota found that caffeine can boost your ability to remember something up to twenty-four hours after it is consumed. Participants received either a placebo or a 200mg caffeine tablet five minutes after studying

a series of images. That is about the same as one or two cups of coffee. The next day all participants were shown another series of images, some of which were the same as before, some similar and some completely new. The caffeine consumers were more likely to correctly identify an image as similar to the ones they had seen, as opposed to

TOP TIPS:

✓ To remember someone's name, say it out loud to another person.

✓ Consuming a cup of coffee after a task can boost your chances of remembering it up to twenty-four hours later.

✓ Doodle whilst listening to something to improve your chances of recalling it later on.

incorrectly identifying it as exactly the same. Coffee, it seems, can sharpen the mind.

So, too, can chewing gum. Researchers behind a study published in the *British Journal of Psychology* asked two groups of twenty participants to listen to a thirty-minute audio recording containing a list of numbers. The group that chewed gum whilst listening recalled the numbers more quickly and more accurately.

Our last piece of advice is simple: doodle. If you are tempted to chastise someone for drawing little stick figures during an important meeting, don't – it aids cognitive performance. A 2010 study by Jackie Andrade and published in the journal *Applied Cognitive Psychology* revolved around playing a boring phone message to a group of forty participants. The twenty who were asked to doodle at the same time remembered twenty-nine per cent more information.

HOW TO SCIENCE YOUR WAY TO SUCCESS IN A JOB INTERVIEW

Interviewing for a new job often ranks among life's more stressful events. Success can mean the difference between your dream job or being stuck in a workplace you hate. Be reassured, then, that scientists have been looking into why employers pick their preferred candidate, and we can bring you the best bits.

You've likely heard that first impressions count, and the research seems to back this up. In fact, according to a 2014 survey of employers, they make up their mind about a candidate within the first seven minutes. It is likely it is even quicker than that – you might have less than a minute. Researchers at the University of Toledo showed thirty volunteers video clips of the first twenty to thirty-two seconds of job interviews and asked them to rate the candidates purely on what they saw. Despite having only seen this fleeting glimpse, the volunteers rated the candidates the same as interviewers who had had twenty minutes with them.

One of the seeds of this success is the initial handshake. A 2008 study, published in the *Journal of Applied Psychology*, found that it was key. Ninety-eight students attended mock interviews where both the quality of their performance and the quality of their handshake were assessed (they weren't told about the handshake evaluation ahead of time). Those who scored highly in the overall interview also had firmer handshakes. A separate study from the same year, published in the prestigious journal *Science*, found that the temperature of your hand is also an important factor. The researchers, led by John Bargh from Yale University, found that participants who had held a cup of coffee were more generous in their evaluation of another person's personality compared to those who had held a cold drink. So, the logic goes, if you warm the interviewer's hand with your own during the shake, they might see you more favourably. Perhaps clutch a cup of coffee while you wait to go in.

It also seems that dogs are not the only ones who sniff each other in order to suss out a new acquaintance. Noam Sobel, from the Weizmann Institute of Science in Israel, believes that we humans do it too – just on the down-low so no-one knows we are doing it. In Sobel's study, 280 participants were filmed meeting with a researcher. Later analysis of the footage showed that for twenty per cent of the time before the meet the participants had their hand close to their face. Curiously, however, this more than doubled once they had shook hands with a researcher of the same sex (*see graph*). Sobel argued that they were subconsciously sniffing their hands for chemical clues about their opposite number. To back this up, he later fitted nasal catheters to participants in a copycat experiment. This showed they were indeed sniffing at the time they brought their hands close to their face. So it seems that making sure your hands are clean is sound advice.

Your appearance holds a lot of currency too, in ways you can and cannot control. According to a CareerBuilder survey of more than two thousand employers on the best colours to wear to an interview, black and blue are the colours to go for – orange outfits came out worst. Employers are also likely to make unconscious decisions based on how attractive you are. That is at least if a 2012 study published in *The Journal of Human Resources* is to be believed. Researchers led by Michèle Belot analysed television game shows that involved contestants voting other players out. They found that less conventionally attractive people were more likely to be eliminated, even if they had performed better than the prettier people. This is part of a well-known phenomenon called the "halo effect". We inadvertently assume that someone who is successful in one area – say attractiveness – is also successful in others. So, dressing to impress might get you some of the way there.

Professional interviewers are supposed to be trained to discount these inherent biases when trying to fill a vacancy, but it is likely they still apply to some extent. Knowing these inside secrets might just be the edge you need to get that perfect job.

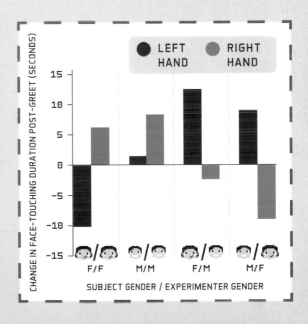

HOW TO STOP PROCRASTINATING

"I'll just check my Facebook page." "One more episode of *Star Trek* won't hurt, right?" "It's alright – I'll just do it tomorrow." Any of these sound familiar? Of course they do. Procrastination affects even the most disciplined amongst us.

A 2007 study by psychologists at the University of Calgary, and published in *Psychological Bulletin*, found that eighty to ninety-five per cent of all college students procrastinated. The rest were probably lying. Studies by psychologist Joseph Ferrari at DePaul University have indicated that up to twenty per cent of us could be chronic procrastinators. In this most serious form, research has found a link to

decreased well-being, poorer mental health, lower performance and financial difficulties.

It seems that we do worse when we have to set our own deadlines, rather than having them imposed upon us by some external party. Back in 2002, a study published in *Psychological Science* looked into this effect, with researchers using their own students as participants. Ninety-nine students

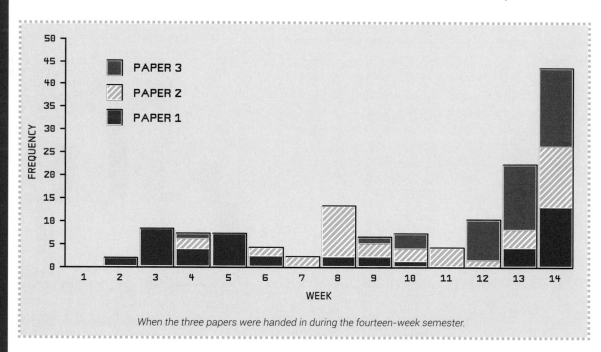

When the three papers were handed in during the fourteen-week semester.

attending a semester-long (fourteen-week) psychology course were split roughly in half. Both groups were told that three essays were due by the end of the semester. However, one group were given imposed deadlines that were equally spaced throughout the course. The others were told they could set their own deadlines. Now, the second group should have been at an advantage. In control of their own time, they could have managed their workload however they liked. And if they chose to submit all three essays towards the end of the semester, they could have benefited from learning more of the course content. Yet the researchers found that not only did the second group largely opt to submit their work way before the end of the course, they got lower marks than the first group.

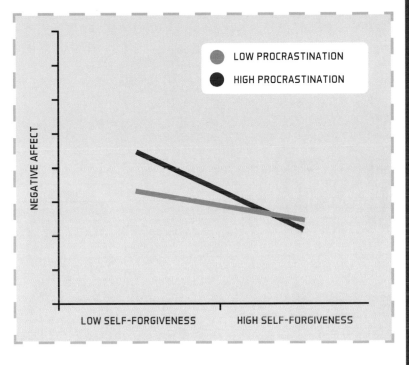

But, never fear, there are scientifically backed tactics you can employ to increase your chances of battling procrastination. First off, give yourself a break – you are only human. We all procrastinate, and berating yourself about it too much has been shown to heighten your chances of procrastinating again in future. Researchers behind a 2010 study, published in the journal *Personality and Individual Differences*, quizzed over a hundred students on self-forgiveness and procrastination just before two exams. Their results show that students who reported high levels of self-forgiveness for procrastinating on studying for the first test procrastinated less in the run-up to the second exam.

Another key tip is to just get started. Many of us can suffer from "analysis paralysis". We spend so much time thinking about how to do something that we overly delay actually doing it. Or, we defer doing a task we are not looking forward to and do something that makes us feel good instead. Hence why you binge-watched that last series of *Doctor Who*.

The biting-the-bullet tactic is backed up by a series of studies by psychologist Timothy Pychyl, a member of the aptly named Procrastination Research Group at Carleton University in Canada. Pychyl intermittently contacted participants and asked them what they were doing, if there was anything else they should be doing, and got them to rate how stressful they perceived each of those tasks to be. He found that participants' perception of their task changed over time. Initially, the avoided task rated highly for stress. But once the dreaded job was actually underway, their stress levels fell. Why stress yourself out more than you have to? Just get on with it.

One way to make getting started a little easier is to break a task down into small chunks that on their own do not seem as intimidating at the whole project seen in the round. Take it from your authors that just writing something, anything, is better than

JUST ONE MORE EPISODE!

staring at a blank page or fretting over how you are going to write thousands of words. You can always edit it later. And it does seem that the very act of doing something is rewarding — even if it doesn't help us actually get the job done any quicker. A 2014 paper by David Rosenbaum from Pennsylvania State University cements this principle.

Participants were given the task of carrying one of two buckets to the end of an alleyway. One bucket was closer to the finish line than the other, and they could choose which one to carry. Surprisingly, the majority of participants picked up the bucket closest to the start line i.e. the one they would have to carry the furthest. The effect was apparent over nine experiments with over 250 students. Rosenbaum quizzed them afterwards and concluded that they probably selected the nearer bucket so they could feel as if they were getting on with the task sooner (even though it required more effort in the long run). He calls this technique "pre-crastinating".

And it seems that humans aren't the only species who pre-crastinate. Rosenbaum later conducted

experiments with pigeons. The birds could earn a reward if they pecked at a screen three times. The first peck had to be inside a box in the centre of the screen, the second either in the same box or in a box that randomly appeared to the side of the centre ,and the third in the side box when a star appeared inside it. So the reward could either be obtained by two centre pecks and one side peck or one centre peck followed by two side pecks. The pigeons were found to plump for the latter, which Rosenbaum says shows they pre-crastinate too — they moved to the end square as soon as possible even though it made no tangible difference to their reward. He later speculated in an article for *Scientific American* that pre-crastination might be common to both humans and pigeons because it appeared in our common ancestors before the two species diverged 300 million years ago.

If all else fails, consider looking at the website stickK. You set yourself a goal and suffer the consequences if you don't meet it. You might offer up $10, for example, which you'll get back if you complete your task on time. Fail, however, and your money gets donated to a cause you really don't like (say the campaign to bring back Jar Jar Binks). According to stickK, their data shows that a financial incentive increases your chances of success by up to three times. If that doesn't set a fire under you then nothing will.

HOW TO NETWORK YOUR WAY TO CAREER SUCCESS

For many people, simply hearing the word "networking" is enough to send a shiver down their spine. The idea of rolling up to a random stranger and striking up a conversation is something that makes a lot of people feel deeply uncomfortable.

Wanting to explore these feelings further, researchers led by Tiziana Casciaro at the University of Toronto conducted a study in which participants were split into two groups. The first were asked to write down a recollection of a time when they had engaged in personal networking (making new friends, asking someone out on a date). The other group recalled a professional networking scenario (one in which the goal was the advancement of their careers). Both groups were then asked to fill in the gaps in the words W _ _ H, S H _ _ E R and S _ _ P. The results, published in the journal *Administrative Science Quarterly*, show that the professional networking group were twice as likely to fill in the words as "wash", "shower" and "soap" as opposed to "wish", "shaker" and "step". Their conclusion was that networking makes us feel dirty.

Perhaps that's a bit of a stretch, but maybe we shouldn't dread networking as much as we do. A 2014 behavioural science study, published in the *Journal of Experimental Psychology*, found that talking to strangers boosts our overall well-being. Not only that, but scientists have found that networking really is a fundamental part of career success. You just have to construct the right web of contacts.

According to a study by Ronald Burt, from the University of Chicago Booth School of Business, it is not the size of your network that counts. Instead, it is how many different networks you are in. We humans have a natural desire to band together. Once someone joins a group, they often put the blinkers on to anything outside that group. People go to the same conferences, hang out with the same colleagues after work, and so on. In short, we get cliquey. The trouble with this approach is that you are likely to hear the same information again and again.

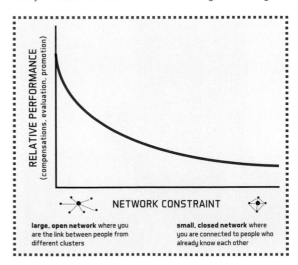

large, open network where you are the link between people from different clusters

small, closed network where you are connected to people who already know each other

In Burt's language, these sort of people belong to a "closed network".

What you really want is to be part of a large open network. The ideal place to be is as a "broker" between many otherwise closed networks. That way, you are pulling a broad range of information from a diverse range of sources. You are also likely to get credit for bringing new information to each group from the outside.

In fact, Burt found that the type of network you are in is the number one predictor of career success (*see* graph). His study found that the extent to which you are a network broker is responsible for half of the differences in salary, evaluation and promotion.

HOW TO BE
MORE PERSUASIVE

We all secretly wish that others would do what we want. And, as you climb up the career ladder, one of the essential skills to develop is the art of persuasion. Particularly when you are managing a team of people or want people to get investors to back your business. Or it could be that you work in sales and you want to persuade more people to buy your product or service.

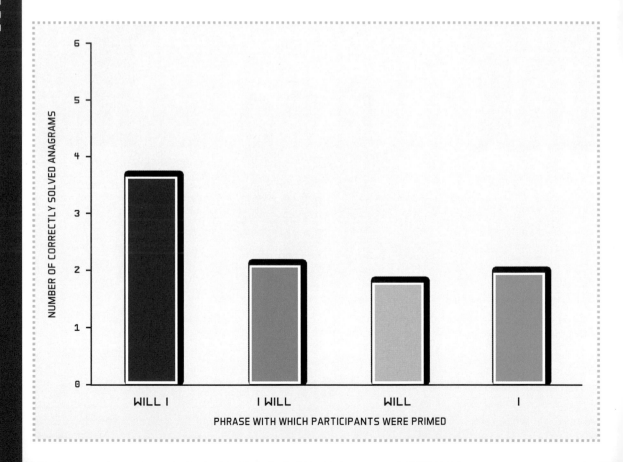

GIVE SOMEONE A REASON TO DO SOMETHING AND THEY ARE **MORE LIKELY TO DO IT.**

Getting someone to acquiesce to your requests is just as much about how you ask as what you are asking. In one study, adding the word "because" boosted compliance by thirty-three per cent. Give someone a reason to do something, and they are more likely to do it. You are also more likely to have success if you *ask* rather than *order*. Perversely, this seems to work, even when we are asking ourselves to do something.

In a 2010 study, published in the journal *Psychological Science*, researchers at the University of Illinois split their participants into four groups before telling them they were going to take part in a handwriting test. Each group was assigned one of four phrases or words to write out twenty times – "I will", "will I", "will" and "I". They were then asked to arrange ten anagrams into new words. Those who had written, "will I" solved twice as many anagrams (*see graph*). This probably seems counter-intuitive, but asking rather than demanding could be more likely to persuade someone to do what you want.

Language really is the secret to persuasion. Dr Robert Cialdini, Professor of Psychology at Arizona State University, has been studying persuasion for decades and his numerous studies back this up. In one, he looked at donations to the American Cancer Society. In the first instance, potential supporters were asked "Would you be willing to help by giving a donation?". The question was then subtly changed to "Would you be willing to help by giving a donation? Every penny will help." Just four words added, but they made a big difference. The amount of people donating jumped from twenty-eight per cent to fifty per cent. The authors concluded that "people are more likely to take action when minimal parameters are set". So, if you want someone to do something, start by giving them a minimum course of action.

Researchers looking for other successful persuasion tactics have turned to Reddit for assistance. This popular website is made up of many forums on a huge variety of topics. One of the message boards is called ChangeMyView, where users post a statement and everyone else tries to persuade them to change their mind. If the original poster is subsequently persuaded, they click a button to bring up the Greek letter delta (Δ) – which as any good geek will know is the symbol mathematicians and physicists use to denote change. They also reveal what it was that made them change their mind. Scientists from Cornell University examined two years' worth of this data and presented their results to the 25th International World Wide Web Conference in 2016.

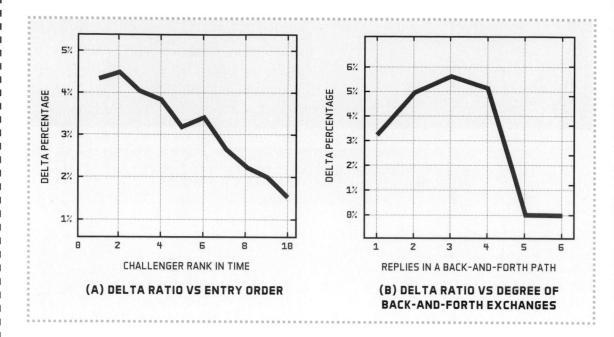

(A) DELTA RATIO VS ENTRY ORDER

(B) DELTA RATIO VS DEGREE OF BACK-AND-FORTH EXCHANGES

The scientists found that the first person to reply was often the most persuasive (*see graph A*). So timing appears crucial – if you want to successfully persuade someone, get in before others have the chance. Also, know when to give up. The rate of success plummeted after several back-and-forth attempts at persuasion (*see graph B*). Longer posts were also more likely to be successful, so don't skimp on detail. And back up that detail with numbers and hard evidence, as this was a surer path to persuasion. Interestingly, the researchers also found that persuasion was more likely if the responder used different language from the original post. Calmer language was also a better bet, with the authors giving "librarian" and "dull" as examples of calm words, whereas "terrorism" and "erection" are words that excite. It is not clear to what extent

these insights extend to face-to-face persuasion, but if you are trying to change someone's mind online or via email, then these are definitely things to keep in mind.

One aspect that is unique to situations in which you can see the other party is body language. And a simple nod of the head can go a long way – yours that is. As social creatures we like to mimic the behaviour of others around us. So the chances are that if you subtly nod your head then so will they – and that's exactly what you want to happen. A 2003 study published in the *Journal of Personality and Social Psychology* found that we are more likely to agree to an argument if we are nodding our head whilst we are listening to it. Pablo Briñol and Richard Petty played two groups of students a piece of audio on the introduction of ID cards under

IF YOU SUBTLY NOD YOUR HEAD THEN SO WILL THEY.

the guise of road-testing a pair of headphones. One group were told to nod whilst listening, the other to shake their heads, both apparently in order to check the headphones worked properly even if the wearer was moving around. Afterwards, both groups were asked questions, both about the headphones and about how they felt about the introduction of ID cards. Those who had been nodding their heads were more likely to agree with what had been said.

So, there is a lot you can do to bring someone around to your way of thinking. On the right is a recap of our top tips.

TOP TIPS:

- ✓ Give someone a reason to do something – say "because".

- ✓ Ask – don't order.

- ✓ Set a minimum course of action.

- ✓ Get in first, back up your argument with evidence and use calm language.

- ✓ Get them nodding along with you.

CHAPTER 3

LOVE AND RELATIONSHIPS

HOW MUCH SEX SHOULD YOU BE HAVING?

It's no secret that most couples love sex - it's the ultimate form of physical intimacy, after all. So you'd be forgiven for thinking that the more sex you're having, the happier you, your partner and your relationship are.

However, that is not actually the case, as research from the University of Toronto clearly shows. Researchers analysed nearly twenty-five years' worth of survey responses collected from over 25,000 US citizens between 1989 and 2012. The surveys asked these participants – all identifying as being "in a relationship" – about how often they had sex and how happy they considered they were.

The researchers' overall finding: sexual happiness has a peak (or, yes, a climax, if you must...). Most couples that had sex once a week were pretty happy, but people's general happiness did not seem to increase with more sex. This was consistent no matter the gender, age or length of relationship. When asked how much sex is "enough", the average answer was five times a month.

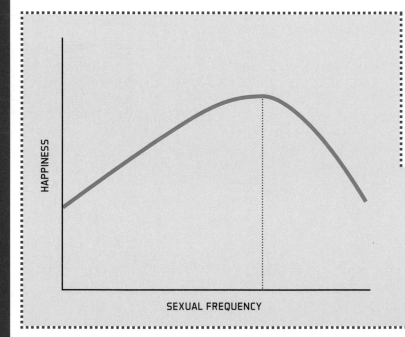

HAPPINESS

SEXUAL FREQUENCY

The relationship between sex and happiness is not linear but curvy-linear – more sex appears to correlate to people's happiness up to a point, but after that point people who had more did not report being any happier in their lives.

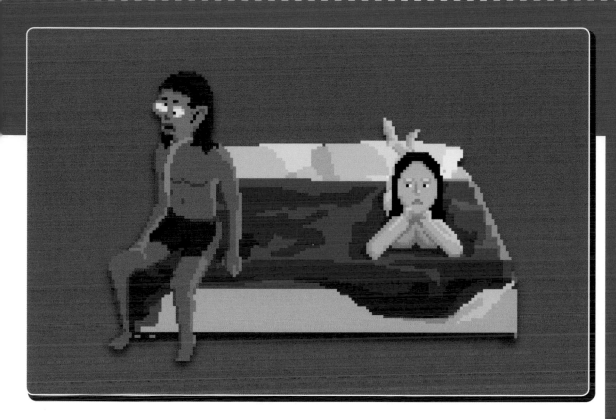

Of course, there is no telling if the person's happiness really is related to the amount of sex they are having, or how satisfied they are with that sex – however much of it there might be. However, there is some back-up in a separate 2015 study that asked thirty-two of sixty-four married couples to double their rate of sexual intercourse for a while. The participating couples managed to have forty per cent more sex on average, but their survey results revealed that they were not any happier. Moreover, they reported feeling less energetic and having worse sex than before.

Why might this be? Well, the Toronto researchers suggest that too much sex "leads to a decline in wanting for, and enjoyment of, sex". In other words, you get a bit bored; when you have too much of a good thing, it takes something of the anticipation, excitement and desire for it away. Additionally, the

Toronto researchers' finding is consistent with studies of other positive non-sexual behaviours (such as socializing with friends), which also found that doing more of it does not uniformly increase a sense of well-being.

If you are wondering how sex compares with money in this kind of analysis, the Toronto researchers also compared people's level of income with their feelings of well-being. People who made less than the average (US$15,000–$25,000 a year) generally seemed to be less happy. Similarly, people who had sex less than once a month were less happy than those hooking up once a week. For more on this topic, *see pages 158–161*.

The take home message is simple: as with so many things in life, it's not how much you're getting, but what you do with it...

HOW TO SPOT A LIAR

Ask any poker player and they'll tell you almost everyone has "tells" that give them away when they are playing. According to experts at the website Pokerology, it's all in the eyes and hands: trembling hands, open, unblinking eyes and full pupils are signs of a strong set of cards; staring into space (or staring deliberately directly at you) means a weak hand.

Some of this theory might work, but the truth is that much of what we believe about lie detection is actually a myth. Psychologist Paul Ekman has tested thousands of people to study their facial expressions and emotions. His research shows that, no matter what it is they are trying to spot, people's accuracy in distinguishing the honest from the dishonest is only around fifty-five per cent – that is just over 50:50 or only slightly better than chance. However, Ekman's research has pointed to one "tell" that seems to consistently bear fruit: micro-expressions – fast, fleeting movements on the face, lasting just a fifteenth or twentieth of a second.

If micro-expressions sound slight and difficult to spot, that's because they are – but they are also hard to control consciously. This, then, leaves something of a double-edged sword. The suspect can't avoid making micro-expressions, but you would be hard-pressed to consciously spot them for yourself. Ekman and his team have conducted experiments with over fifteen thousand people and only about fifty of those – less than 0.3 per cent – seemed to pick up on them with any regularity.

Yet, the fact that as many as fifty individuals were able to spot the micro-expressions at all is intriguing. Could such unconscious signals, in some cases, be the source of our "hunches" or intuitions about liars? That feeling that your date isn't going to call you back like they said they would; that the skinny naked guy isn't really going to lead you all the way to Mordor...?!

This concept was put to the test by researchers at the University of California, Berkeley. They showed participants videos of suspects being grilled about stealing US$100. Half of the suspects were actually telling the truth. Each volunteer was shown a video of one of these honest people as well as a video featuring a liar. Later, the participants in the experiment were set a task – pressing buttons to categorize words such as "honest", "genuine" or "deceitful" as either associated with telling the truth or lying. The catch was that these words appeared alongside photos of the suspects that they had seen in the videos (remember, the volunteers didn't know if *anyone* was telling the truth), though the participants were asked to ignore the photos and concentrate just on the meaning of the word by itself.

It turns out that people were faster at correctly categorizing words as truth or lie when they were shown alongside the photo of the actual liar or truth-teller. This suggests that on some unconscious level, the concept of "liar" or "honest person" was put into the participant's mind, swaying their thinking – even if they weren't aware of it.

Other studies have found that a person's ability to accurately detect a cheater can be improved if they are distracted (in this case by doing a demanding puzzle at the same time), while passing their judgement. If pressed for an immediate decision, they did no better than chance – which, the University of Mannheim psychologists who did the experiments say, shows that our brains have the capacity to distinguish between truth and deception. However, this requires integrating a rich set of cues, some of which are very subtle and of which we may not even be consciously aware.

So, can you spot a liar? While the indications are to follow your instinct and distract yourself so that you don't think about it too hard, beware. For every unconscious cue, there is also unconscious bias. Our brains have a proven tendency to stereotype people based on their age, race or background as a quick reference, a form of shorthand. We therefore tend to favour people who are similar to ourselves. Lie detection is never one hundred per cent certain, but if there is one thing we know for sure, it is that our own judgement is fallible.

WHAT SCIENCE KNOWS ABOUT SUCCESSFUL MARRIAGES

HEY, SIMILAR SPENDER

In a paper jovially entitled "Fatal (Fiscal) Attraction: Spendthrifts and Tightwads in Marriage", Scott Rick and colleagues at the University of Michigan found that, when it comes to money, it is true that opposites attract.

Surveying hundreds of married couples, the researchers found that significantly more people were partnered up with those who had a very different spending behaviour to their own. However, follow-up studies showed that these couples also suffered more arguments over money and reported lower satisfaction with their marriages than those who were married to people with similar spending habits to their own.

Of the mixes of couples, spendthrifts who married tightwads seemed to come out best – at least in the financial sense – no doubt thanks to their partner putting a brake on their spending. However, in terms of relationship satisfaction, this group was beaten by spendthrifts who married other spendthrifts. This finding possibly demonstrates that money can, in a sense, buy happiness (albeit along with larger debt). However, any real smugness in this experiment belonged to the tightwad couples – who were generally content with their lot, and had a better bank balance.

WORDS MATTER

When your partner does something for you, do you say, "Thank you"? Or, like many who took part in a 2007 survey set by Jess Alberts and Angela Trethewey at Arizona State University, do you just assume that he or she "knows" that you are grateful to them?

Those two little words of gratitude matter, Alberts and Trethewey found. Whether it is married couples or roommates, when it comes to crucial things like the division of housework, an expression of gratitude from the recipient of kindness is often sorely missing. In a follow-up, the researchers also found that (shock, horror) those who felt appreciated harboured less resentment and had more satisfaction with their relationships than other participants.

If none of this sounds like rocket science, then neither will this advice: try to make sure partners "own" the tasks they are set, or switch tasks and lists regularly. If your partner is under-performing, don't wait for your tolerance to break, but instead explain what you would like done and when. And for goodness' sake, when they finally do something for you, say thank you...

YOU ARE ANNOYING – AND THAT'S OK

Who is the most irritating: your partner, kids or friends? Partners, say a majority of eight hundred people surveyed by the University of Michigan's Institute for Social Research. The bad news is that our negative views of our spouses seem only to increase over time.

The good news is that relationships can thrive on negative behaviours – arguments, blaming the other, nagging your partner to change – according to a decade of separate research led by psychologist James McNulty at the University of Tennessee. However, it all depends on your expectations.

Having positive expectations only helps if those expectations are met, according to one of

McNulty's studies of newlyweds. However, couples with more problems do better if they expect challenges. In a separate study, McNulty found that couples who reported fewer problems wrote off negative behaviours as "outside of their partner's control", and seemed more satisfied with their marriage as a result. However, if a couple had *more* problems, actually blaming their partner for their actions correlated with higher marriage satisfaction. According to McNulty, if your partner is far from a saint, it's better not to look the other way. This can help motivate partners to change in the long run, even if it doesn't feel particularly good at the time.

THE BEST TIME, MATHEMATICALLY, TO SETTLE DOWN

Over the years, mathematicians have called it a lot of things from "the secretary problem" to "the sultan's dowry problem". You might also call it the "grass is always greener" problem - and apply it to choosing houses or chairs as much as to potential life partners. This is because it essentially comes down to the same issue: what are the chances that, if I wait a bit longer, something or someone better might come along?

Let's break it down. Essentially you are choosing from a set number of options — you just don't know how many. But let's say, hypothetically, that there are eleven people in your lifetime who could be "the one" for you. They're all nice. The trouble is you meet them one by one (usually!), over however many years. And they breeze into your life in random order.

This state of affairs offers no scope for like-for-like comparison. How could you possibly see who's best, let alone whether the next person along might be better? You can't. So you have to take a chance...

However, here's a magic number that can dramatically increase your chances of relationship success: thirty-seven per cent (or 36.8, if you want to

NO PARTNER

WRONG PARTNER

RIGHT PARTNER

be really precise). That's the calculated percentage of partners you should date and reject before you decide to settle down – which, on average, works out to about four people in a typical lifetime. At that point, pick the next one that seems better than any of your previous boyfriends or girlfriends. According to maths, you stand a greater chance than at any other point of that next person being "the one".

I wouldn't blame you for feeling sceptical about this revelation. In practical terms, it's a bit of a stretch. Not only can you never know how many options (partners) you might potentially have – but what counts as a "potential suitor" as opposed to a "fling"? How exactly do you define your "best match/perfect partner/bestest love in the world evah" anyway?

But here's another number for you: nine per

$$P(r) = \frac{r-1}{n} \sum_{i=r}^{n} \frac{1}{i-1}$$

37% THE PERCENTAGE OF PARTNERS YOU SHOULD DATE AND REJECT BEFORE YOU SETTLE DOWN.

cent. That is the probability of you picking the best of eleven potentially serious partners if you were to pick randomly. Use our crazy plan and the odds rise significantly to – yes – thirty-seven per cent.

Still, thirty-seven per cent is not one hundred per cent. You run the risk that one of the *first* thirty-seven per cent of partners that you date might be the best fit. But the experts have done the maths: if maximizing the chances of finding, meeting and sticking with the best of the bunch is what matters to you, then statistically this strategy has the highest chances of you achieving success.

And when you think about it, it makes sense. In fact, it's actually kind of obvious. All you are doing, really, is balancing out your risk of choosing too late or too soon. It allows you enough dating experience to find out what your tastes are – what kind of person you like and what kind you don't – and avoids you deliberating so long that you run out of options.

It makes even more sense if you imagine the whole thing as a game (in which winning = settling down with the best match), and then map out the odds of you winning as the number of potential partners increases from one upwards.

Let's say my tactic is to pick the first person that comes my way. If I only ever have one potential "the one" to choose from, I would win every time. If I have two potential candidates, there's a fifty-fifty chance of me picking the best one – no matter if I pick at random or try to strategize.

With three potential life partners to choose from, things become more complicated, but this is where our crazy tactic starts to come into its own (*see diagram*).

The pattern maps out as you factor in bigger and bigger groups – no matter the range, the optimal number of options to "try before you buy" always comes back to thirty-seven per cent.

Want to better the odds? Then lower your standards... Let's say you don't necessarily need "the best" partner, but you would be happy with just a "good" partner. In this case, your magic number is thirty per cent, which is great if time is of the essence to you – it amounts to fewer people to date, after all. Using this strategy, if your total number of potential good matches was ten, you would succeed in finding them three out of four times. If your number of potential good matches was one hundred people, you would be ninety per cent certain to choose someone that you would be happy with, which sounds like pretty good odds to me.

But what if you don't want to settle? What about those perfectionists, for whom it's the best or nothing at all – people who'd be perfectly happy staying single otherwise? Tweak the formula again, as Japanese mathematician Minoru Sakaguchi did in 1984. The magic number he came up with was 60.7 per cent of potentials. That's a lot more people to review – requiring more effort and a good deal more time. But if you're comfortable with the idea of being single, then what's a few more years to ensure you find – definitively – the Neo to your Trinity, the Sheldon to your Amy? For these people, they're better off playing the field for longer.

Whatever your preference, this magic number potentially makes FOMO less of a factor in your search for eternal happiness...

PLAYING THE ODDS TO "WIN" AT LONG-TERM RELATIONSHIPS

This diagram compares the probability of success if you were randomly selecting amongst three partners. Each box represents a partner and the number ranks their quality (1 = the best). Following the tactic of rejecting the first thirty-seven per cent and picking the next (in this case, rejecting the first two and choosing the third) has you "winning" more than any other tactic.

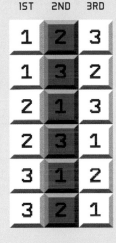

1ST 2ND 3RD

1	2	3
1	3	2
2	1	3
2	3	1
3	1	2
3	2	1

If you choose the first suitor you date every time, you will **win twice**

(2/6)

1ST 2ND 3RD

1	2	3
1	3	2
2	1	3
2	3	1
3	1	2
3	2	1

If you choose the second suitor you date every time, you will **win twice**

(2/6)

1ST 2ND 3RD

1	2	3
1	3	2
2	1	3
2	3	1
3	1	2
3	2	1

If you choose the third suitor you date every time, you will **win twice**

(2/6)

1ST 2ND 3RD

1	2	3
1	3	2
2	1	3
2	3	1
3	1	2
3	2	1

BUT if you date and reject the first suitor, then choose the next suitor who is better than the first, you **win three times**

(3/6)

HOW TO DEAL WITH A BREAK-UP

As a million love songs tell us, breaking up is hard to do. And it can hurt - literally. MRI brain scans of the heartbroken have shown similar patterns to those experiencing withdrawal from cocaine. Brain areas associated with physical pain appear to be activated in emotional pain, as well.

A bad break-up could also disrupt your entire physiology: studies have found that long-term couples seem to share similar biological rhythms in terms of sleep, appetite, body temperature and heart rate. So, what tips can science offer to help heal the pain of break-up?

REMEMBER WHO YOU ARE

The more time you spend with a significant other, the more your sense of self becomes entwined with theirs, with a break-up naturally disrupting each person's sense of who they are — at least, according to a 2010 paper in *Personality and Social Psychology Bulletin*. By studying six months of surveys and personal diary entries, researchers found that participants were more likely to use words such as "confuse" and "bewilder" than people who had not been through heartache. According to the researchers, the less clear participants felt about their sense of self, the more emotionally distressed they seemed. So, in times of emotional heartache, remind yourself of the things you love — immerse yourself in your hobbies, hang out with your own friends.

THINK ABOUT IT

Although it is clearly not a good idea to wallow in self-pity, taking some time to reflect on your situation can

help. In a 2015 study published in the journal *Social Psychology and Personality Science*, 210 volunteers who said they had recently been heartbroken took part in surveys over a period of nine weeks. Half of them answered questions about their break-up at various points throughout the study period, but the other half just did general surveys at the beginning and end of the study. According to the data, those who talked progressively about their experiences developed a stronger sense of who they were and seemed to feel less lonely as a result of their break-up.

DON'T CYBERSTALK

It sounds obvious, but staking out your ex's Facebook page after a break-up is really not a good idea. If you need evidence, look at the results of a 2012 study conducted by Dr Tara C. Marshall at Brunel University: 464 surveyed women who did so were more likely to feel distressed, negative and pining for their ex, while also being less likely to experience personal growth and "move on".

Mind you, burning all your bridges isn't wise either. The study also found that those who "un-friended" their exes suffered just as much, perhaps due to the alluring uncertainty of not knowing what they were up to. The researchers suggest that inoculation with the occasional banal status update may help reduce the ex's appeal.

YOU WILL GET OVER IT – AND FASTER THAN YOU'D THINK

Observing seventy volunteers and asking them each week about their relationships, participants in a 2008 study published in the *Journal of Experimental Social Psychology* anticipated that on average it would take twenty weeks to bounce back from a break-up. However, of the twenty-six volunteers who did actually break up, it took only ten weeks for them to feel better. Moreover, their sense of distress was much lower than the level they had predicted they would feel.

EAT CHOCOLATE

Seriously. Chocolate triggers a surge in opiates, the same neurotransmitters that ease physical pain.

TOP TIP:

Facebook has tools that allow you to temporarily block or "un-follow" certain friends. Use these tools for a much needed "time out", without impulsively erasing those friends completely from your life.

HOW TO SUCCEED ON TINDER (AND OTHER DATING SITES)

It is tougher than ever to meet new people, let alone find "the one". And since the Internet has become central to our lives in so many ways, it is no surprise that online dating dominates the way that people look for partners today.

Thankfully, a 2012 study conducted by Stanford University researchers found no difference in the quality of online or offline-started relationships (although there are enough other conflicting studies to cause confusion). However, if online dating really is the new reality, let's delve into the data and find an evidence base for how to do it right.

PHOTO(DA)BOMB

Whether you are using Tinder or one of its competitors, put more time and effort into your photos than anything else. Users of dating sites say they draw more information about someone's personality from a photo than from answers to random questions posted on the website. In actual studies, any characteristics that participants previously said were of value in a potential partner seemed to go out of the window once they met them in real life. This would appear to indicate that looks often matter a lot more than other factors.

One tip to successful online dating is to open up – literally. A 2016 study of videos from a speed-dating event found that those displaying "postural expansiveness" – leaning back or spreading their arms and legs wide, literally expanding the body in physical space – were more likely to gain romantic

not slouching
+
pointing your feet towards the camera
+
look neat and composed
+
smile
=
extrovert, healthy, energetic

♡ ♡ ♡

slouching
+
dishevelled
+
frowning/straight
–
face
=
brooding, introvert

interest. Such people were perceived as dominant and open, compared to others who were hunched or crossed their arms or legs, the University of California researchers wrote in the Proceedings of the National Academy of Sciences.

You needn't necessarily try too hard, though – especially if you are a woman looking to attract the opposite sex. According to a University of Connecticut study, enhanced photos increase attractiveness but lower trustworthiness among men. This finding was based on a study of 305 women, using pictures of them taken either in their normal everyday state or enhanced by using flattering lighting, hair care and make-up. Meanwhile, female participants in the experiment said that they found the enhanced photographs of men increased both factors.

SHOW OFF YOUR FRIENDS

A solo photograph is important for your main dating website image – no one wants to play the guessing game of which is you in a picture, especially during the split-second, quick-swipe environment of Tinder. However, for your other piccies, show off your X-Men (or women). Analysis of Tinder data indicates that a sign of an outgoing personality (favoured by many, but not necessarily all), is the presence of a group in a photograph rather than just a lone Wolverine.

PICK A USERNAME NEAR THE FRONT OF THE ALPHABET

Although a review of eighty-six studies about online dating reported that "a variety of measures of success [in the

offline world] ... are correlated with names higher up in the alphabet", there is one obvious reason to go with, for example, "ADude" rather than "TheDude". This is that screen-names starting with a letter near the top of the alphabet are presented first in search listings, so those towards the end of the alphabet are often lost at the bottom of the pile. Sure, people can sort and filter if they want to be thorough, but don't underestimate how lazy people are, especially when they are essentially looking at things at random.

As to whether it is better, if given the choice, to go for a shorter nickname or your full name – it depends. Although by no means representative of the whole population (it's just one app and one pool of users, after all), in March 2016 the dating app The Grade published statistics on their most right-swiped (that is, selected) names.

MEN		WOMEN	
MICHAEL	12.7%	REBECCA	59.7%
MIKE	12.6%	BECKY	22.5%
DAVE	18.6%	JEN	54.3%
DAVID	13.4%	JENNIFER	44.9%
MATTHEW	16.9%	ELIZABETH	58.9%
MATT	15.4%	LIZ	47.6%
STEVE	13.2%	ALY	59.0%
STEVEN	12.6%	ALISON	57.5%
STEPHEN	11.7%	ALI	51.5%
		ALLIE	50.4%
RICK	17.1%	KATIE	60.8%
RICHARD	7.0%	KATHLEEN	59.0%
RICKY	15.5%	KAT	47.1%

KISS ("KEEP IT SIMPLE, STUPID")

"People are naturally drawn to words that are easy to remember and pronounce," wrote Khalid S. Khan of Barts and the London School of Medicine and Sameer Chaudhry of the University of North Texas in "An evidence-based approach to an ancient pursuit: systematic review on converting online contact into a first date", which was published in 2015 in the journal *Evidence-Based Medicine*. Simple language is easier to understand, which also increases likeability. They add that "Overall attractiveness of the text is positively correlated with photo attractiveness," so if your photo succeeds in getting someone to stop and read your profile message, an appealing text will reciprocally increase the exposure time to your primary photo. This, the researchers concluded, can increase their liking.

Of course, this is easier said than done. Too many people struggle with what to write and end up waffling on aimlessly. Keep it snappy, and don't just talk about yourself. Aim for a 70:30 ratio of who you are to what you are looking for in a partner. A profile that is all about you, propose Khan and Chaudhry, "will attract far fewer responses than a combination of who you are and what you are looking for".

YOU DON'T HAVE TO BE A MODEL...

...but you're in luck if you're a pilot or physiotherapist. According to Tinder's published stats, male pilots and female physical therapists top the charts for swiped professions (at least as far as US users are concerned). The male professions that received the most right swipes among female users were the rather stereotypical fire fighters, doctors and television/radio personalities. Having said that, in at No. 2, just behind pilots, say hello entrepreneurs... They also ranked highly (third) among the right-swipes of men, alongside PR workers and teachers, with the aforementioned physios and interior designers taking the top two places for attractive female professions. Yet, surprisingly for a service that places so much emphasis on photos, models ranked in the lower levels of the top ten for both genders — the Derek Zoolanders of this world ranking number eight, and the Giselles even lower at number ten.

HOW TO MAKE FRIENDS THROUGH KARAOKE AND DANCING

Maybe you love them, maybe you hate them – but science has demonstrated why singing and dancing are key components of our social culture.

Music – either listening to it or creating it – can reduce aggression, improve mood and generally make people more cooperative. Bonding over a shared love (or hate) of hipster Indie rock or death metal is, after all, one of the most common ways of making friends. One study has even shown that out of one hundred evening-class students surveyed, those in a weekly two-hour singing class reported feeling closer to their classmates than others who had taken art or creative writing instead. After a month of classes, on a scale of 1–7 for general camaraderie (7 being the best), the singers felt a whole two points closer to their group since the beginning, compared to just a half point in the art or writing classes. However, it is important to note that the difference between classes was less after seven months, when – as is the nature of evening school – all the classes had shrunk due to dropouts, perhaps allowing for people to get to know each other better.

The researchers speculated that singing as part of a group participation activity bonds everyone together. This approach obviates the need for the usual "one-on-one" interactions that groups need to break the ice and make individual connections. When you sing as part of a group, you have a shared goal – and the whole point is to do something in sync. Moreover, singing has been shown to release hormones – such as endorphins and oxytocin – that make us feel happy and relaxed and enhance feelings of trust.

Dancing may have similar effects. Like singing, it is thought to have been an important part of our tribal ancestral societies. And not only has it been shown to form stronger social connections, dancing in sync can actually raise a group's pain threshold. Researchers from Oxford University asked 264 students from a Brazilian island to learn four dance moves. They then made them perform the moves either in our out of sync. Before and after the dance-off, the scientists recorded the islanders' blood pressure, but in putting the measurement cuffs on, inflated them until the students said it felt uncomfortable. Intriguingly, those who had danced in sync tolerated an additional twenty units of air pressure in the cuffs after dancing, compared to their out-of-sync peers. Even those who had followed the dance moves while sitting down (though still in sync), showed a similar drop in pain sensitivity, not to mention reporting a generally increased feeling of camaraderie with their fellow dancers. The findings of this study match those of many previous studies, which have similarly shown how tapping in sync, or even walking in step, can lead people to trust one another more and be more willing to help each other.

HOW TO DO A BOOGIE, NOT A BOGEY

MOVE THIS

AND THIS

BUT THESE...
NOT SO MUCH

DO move your head and torso in time to the groove **#ShowMeYourMoves**

DON'T flail your arms and legs about **#ShowMeYourMoves**

DO vary your moves **#ShowMeYourMoves**

DON'T just do the robot for the whole night **#ShowMeYourMoves**

Using motion-capture, scientists at Northumbria University made computer avatars of nineteen men who were asked to dance to a drum beat. Thirty-seven women of the same age group then ranked the plain-looking avatars in terms of skill. The results indicated that varied movements of the head, neck, upper body and right knee were associated with high scores from the ladies.

WHEN TO
TEXT NEXT

Here's a fact for you: in 2010 only ten per cent of young adults used texts to ask someone out for the first time. By 2013, that figure had risen to thirty-two per cent. That's a third of all people reportedly taking their first tentative steps towards romance by text message.

Yet for all the benefits of doing so remotely via text/email/WhatsApp/Minecraft, this approach also involves the waiting game. What you need is a walkthrough. The unlikely source is actor and comedian Aziz Ansari – you might remember him from such geek TV favourites as *Parks and Recreation*. After one rejection too many, he teamed up with New York University sociologist Eric Klinenberg to take a scientific approach to dating in their 2015 book *Modern Romance: An Investigation*.

I'm willing to bet that when you get a text message from the object of your affection, your first instinct is to reply straight away – you don't want to anger them by keeping them waiting, right?

Wrong. Numerous studies conducted over the years on rats, mice – and, indeed, humans – show how much anticipation raises our perceived value of a reward. For example, rats that receive a treat by pushing a lever become, over time, less likely to rush to that lever. They know it will always be there. Not that you're a rat, of course, but psychologists say the same principle applies to human beings. It is what they call the "scarcity principle": if we feel that something is always going to be there, we take it for granted, lowering its value as a reward. However, if we don't know whether it's going to be there or not, it's more attractive – causing a mixture of anxiety

and excitement in our brains. "Reward uncertainty", as the scientists call this syndrome, can dramatically enhance levels of dopamine – the "happy hormone".

Another thing that affects all this is that our behaviour in messaging is very different to that which applies to other methods of communication.

In the days when we only had landline telephones, people wouldn't bat an eyelid over waiting for a reply — even when playing voicemail tennis over several days. However, now that we all have smartphones practically glued to the palms of our hands, we expect an instant response. Now consider what scientists call "habituation" — the diminishing of an innate response to a frequently repeated stimulus. In other words, we just become used to the particular stimulus to the extent that it is not so much of a rush when it happens. However, it is very noticeable when something breaks that pattern.

In Anzari and Klinenberg's surveys, people said they would wait anywhere between ten and sixty minutes to respond to a text, depending on the previous pattern of communication and, perhaps crucially, the state of their relationship. When the person you are messaging is not so well known to you, you have no frame of reference for how things will go. But our habituation to messaging is naturally for that "quick hit" of replies. Dr Natasha Schüll, an anthropologist who studies gambling addiction at MIT, has drawn parallels with playing a slot machine. As she told *Nautilus* magazine, "There's a lot of uncertainty, anticipation, and anxiety. Your whole system is primed to receive a message back. You want it — you need it — right away, and if it

doesn't come, your whole system is like, 'Aaaaah!' You don't know what to do with the lack of response, the unresolved outcome."

The "hit" that derives from that uncertainty you feel when you are waiting for a text can form a strong basis for attraction. In a 2011 study published in the journal *Psychological Science*, researchers showed forty-seven women the Facebook profile pages of men who, they were told, had rated the women's profiles as " best", "average" or "uncertain". Astonishingly, the women in the study said that they were most attracted to the guys whose opinion of them was "uncertain" — even more so than those who had liked them the best. The women went on to say that they had thought more about those guys over the next few days than any of the others.

Times have changed. In days gone by leaving a message on an answer phone was like buying a lottery ticket, knowing you needed to wait for the results. As Schüll puts it, "You weren't expecting an instant callback and you could even enjoy that suspense, because you knew it would take a few days. But with texting, if you don't hear back in even fifteen minutes, you can freak out."

Texting and messaging has changed us. But it needn't dictate your romance.

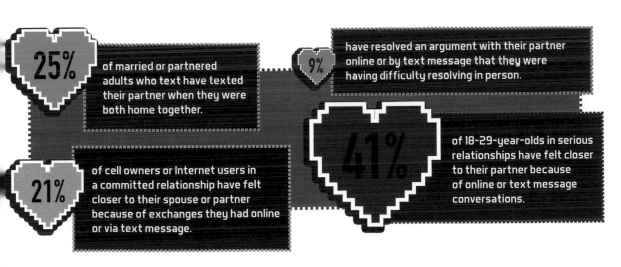

25% of married or partnered adults who text have texted their partner when they were both home together.

9% have resolved an argument with their partner online or by text message that they were having difficulty resolving in person.

21% of cell owners or Internet users in a committed relationship have felt closer to their spouse or partner because of exchanges they had online or via text message.

41% of 18-29-year-olds in serious relationships have felt closer to their partner because of online or text message conversations.

CHAPTER 4

AT HOME

WHAT'S THE BEST WAY TO GET KETCHUP OUT OF THE BOTTLE?

A bowl of fries or a hot dog can be instantly improved with the addition of ketchup. But if you're not careful, you can end up ruining it, too.

We've all been there — no matter how hard you try, you just can't get that stubborn sauce to flow out of the bottle. Worse still, you manage to get it out, but instead of neatly delivering a perfect, velvety dollop of the red stuff, you drench your food in a stream of vinegary red water. It's a fine line, but fear not — physics can ensure that you get the perfect pour every time.

The secret lies in the fact that ketchup doesn't behave like most liquids. It is what physicists refer to as a "non-Newtonian fluid". In the seventeenth century, the famous British physicist Sir Isaac Newton worked out the rules behind how forces affect ordinary liquids such as water. Yet there are some renegade materials that don't follow these rules, ketchup being a perfect example. Such materials are basically "two-faced": they can behave either like solids or as liquids, depending on how you treat them. Force ketchup into being a solid and you don't stand a chance of liberating it from the bottle. Coerce it into being overly liquid-like, however, and your fries are sure to be drowned.

The key is how the particles within the ketchup behave. Think of them as tiny pool balls. When the ketchup is resting on the table, the particles are all crammed together and cannot flow past each other to get out of the bottle. Hitting or shaking the

bottle forces these particles to change shape and become stretched out, meaning that they can flow by one another more easily. But here's the kicker with non-Newtonian fluids: shake too vigorously and the ketchup's viscosity changes dramatically and it becomes one thousand times thinner. Cue your fries bobbing in a red sea...

So, whatever you do, don't shake the life out of the ketchup bottle. There are two main scientifically backed tactics you can employ to conquer the ketchup conundrum. Either apply a lot of force over a short amount of time, or apply a small force for a long amount of time. But never apply a lot of force for a long time.

Here at Geek Towers we've put in the hours and road-tested the best ways to exploit ketchup's non-Newtonian properties. In the bronze medal position we have long, slow shakes of the bottle with the lid off. This slowly, slowly approach invigorates the ketchup particles sufficiently to flow, but be warned that it is fraught with danger. If you are shaking with the lid off, this can spell disaster if you misjudge it. Nabbing the silver medal is the good old short, sharp shunt on the end of the bottle. But the runaway winner is to turn the bottle upside down whilst keeping the lid on, followed by just a couple of strong shakes. Remove the lid and the ketchup flows like a dream.

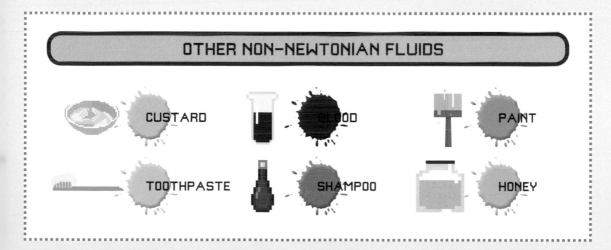

OTHER NON-NEWTONIAN FLUIDS

CUSTARD

BLOOD

PAINT

TOOTHPASTE

SHAMPOO

HONEY

HOW TO BOIL THE PERFECT EGG

How do you like your eggs in the morning? Soft boiling them is certainly one of the easiest methods and your reward is the ability to dip some bread into the velvety, runny yolk. But if you've ever boiled an egg, then you'll surely have noticed one fundamental drawback to this way of cooking: it is very hard to get both the white of the egg and the yolk cooked to perfection.

When boiling an egg, often you'll hone in on getting the yolk bang on, but end up with a flabby, undercooked white around it. The reason for this difference lies in the different chemical make-up of the two parts of the egg.

Inspired by this longstanding problem, scientists have set their minds to tackling the question of how best to boil an egg. One man who has examined the issue in considerable detail is physicist Charles Williams at the University of Exeter in the UK. His work has led to the following formula for the optimum boiling time:

$$T = 0.451 M^{2/3} \ln \left[0.76 \times \frac{(T_{egg} - T_{water})}{(T_{yolk} - T_{water})} \right]$$

Don't worry if you don't understand the maths — we'll explain it all for you shortly. The equation simply describes how the overall cooking time in minutes (T) depends on the mass of the egg in grams (M), the initial temperature of the egg (T_{egg}) and the desired temperature of the yolk (T_{yolk}). We'll assume as we're boiling the egg that the temperature of the water (T_{water}) is 100°C (212°F).

So, just what is the ideal temperature of a perfectly cooked yolk? It turns out the answer is between 63 and 65°C (145 and 149°F). Go above this temperature and the yolk starts to thicken to the consistency of treacle — not ideal for dipping that bread. Let's assume your egg came straight from the fridge (so T_{egg} is 4°C/39°F) and we'll put 63°C (145°F) into Williams' formula as T_{yolk}. Doing

BOILING AN EGG UP MOUNT EVEREST

We've said that one of the issues with getting the perfect boiled egg is that water boils at 100°C (212°F), yet that's not the ideal temperature for either the yolk or the white to cook. Using a water bath to regulate the temperature might be one solution, but you could try climbing a mountain to boil your egg instead.

That's because water only boils at 100°C (212°F) at sea level. If you increase your altitude then you are also decreasing the amount of air pressing down on the water molecules. This means it is easier to break the bonds between them and therefore for the water to boil. Go just 500 metres (1,640 feet) up and water will boil at 99.5°C (203.9°F). Get to 3000 metres (9,843 feet, and the height of The Fortress in the Canadian Rockies) and you'll have broken the 90°C (194°F) barrier.

By the time you get to the summit of Mount Everest at 8,848 metres (29,029 feet) you can get your egg on a rolling boil at just 70°C (158°F). As it is boiling it won't get any hotter than this. This is almost the ideal temperature for a velvety yolk, but alas is not hot enough to fully cook your white. Instead you'll have to resort to the water bath technique of cooking the perfect yolk and finishing the white off in a pan. But what better way to celebrate standing on top of the world?

so tells us the following: a small egg (47g/1.6oz) will take four minutes; a medium egg (57g/2oz) four and a half minutes; and a large egg (67g/2.3oz) will take five minutes. For an egg initially at room temperature (20°C/68°F) this shortens to 2.9 minutes for small, 3.3 minutes for medium and 3.7 minutes for large.

So, if you want to geek up your egg cooking, that is a good place to start. However, you don't really want boiling water involved at all. The yolk may cook perfectly at 63–65°C (145–149°F), but its different chemical structure means that the white cooks best at around 82°C (179°F). Forcing them both to bathe in boiling water isn't helping on either front.

According to a 2011 study in the journal *Food Biophysics* entitled "Culinary Biophysics: on the Nature of the 6X°C Egg", it isn't just temperature that is key for yolks, but time as well. The research by César Vega and Ruben Mercadé-Prieto suggests cooking the egg in a water bath where the temperature is constant and in the range between 60 and 66°C (140 and 150°F). You can then vary the time the egg is cooking for depending on how runny you like your yolk to be (*see table and graph overleaf*).

63–65°C THE IDEAL TEMPERATURE OF A PERFECTLY COOKED EGG YOLK.

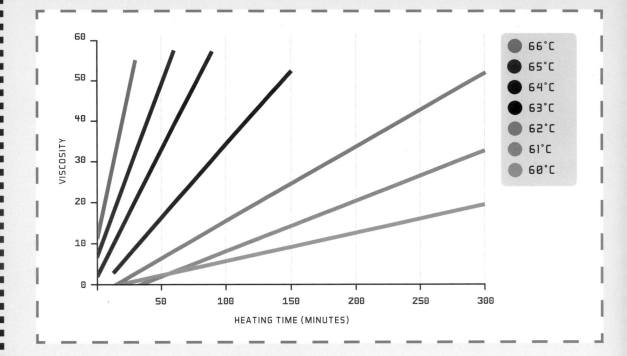

Legend:
- 66°C
- 65°C
- 64°C
- 63°C
- 62°C
- 61°C
- 60°C

X-axis: HEATING TIME (MINUTES) — 50, 100, 150, 200, 250, 300
Y-axis: VISCOSITY — 0, 10, 20, 30, 40, 50, 60

FOOD	VISCOSITY	
WHIPPING CREAM	0.02	
RAW EGG YOLK	0.09	
PANCAKE SYRUP	0.96	
CHOCOLATE SYRUP	1.4	
SOUR CREAM (17% FAT)	2.9	
GREEK-STYLE YOGHURT	3.0	
MOLASSES	3.3	
SWEETENED CONDENSED MILK	6.8	
MAYONNAISE	12.1	
READY-TO-EAT CHOCOLATE PUDDING	13.8	
HONEY	18.3	
NUTELLA®	28.1	
COOKIE ICING (FRESH)	29.3	
TOOTHPASTE	43.8	
MARMITE	43.9	

DON'T HAVE A WATER BATH? PLUNGE YOUR EGG INTO BOILING WATER FOR 30 SECONDS AND THEN ADD ICE.

For example, cooking the egg at 63°C (145°F) for around an hour will result in a yolk that is slightly more viscous than honey. You then remove the shell and partially cooked white from around the yolk and finish cooking the white in a frying pan, before serving it on top of the yolk.

The results from this method are phenomenal and you'll have a hard time finding a more delicious egg than this. However, we realize that not everyone has a water bath, wants to wait an hour for their egg,

or wants their white separate from their yolk.

If this is you then you could turn instead to a shortcut devised by chef J. Jenji Lopez-Alt in his book *The Food Lab: Better Homecooking through Science*. He suggests that the best way to boil an egg is to plunge it into boiling water for thirty seconds and to then add ice to lower the temperature to 82°C (179°F), before turning the heat right down to maintain this temperature as closely as possible. Remove the eggs after six minutes for soft-boiled or eleven minutes for hard.

SHOULD YOU TRUST THE "FIVE-SECOND RULE"?

C'mon, we've all done it. You're preparing some food and you accidentally drop some of it on the floor. Acting swiftly, you retrieve and eat it, before invoking the famous "Five-Second Rule" to back up your case to any disgusted onlookers.

The five-second rule is the belief that any food picked up off the floor within that timeframe will not have enough time to get "dirty" and so will still be safe to eat. This urban legend varies slightly in length around the world, with some claiming shorter or longer durations, but is it a myth or something that is supported by scientific research?

US high school student Jillian Clarke first investigated the issue back in 2003, while she was volunteering in a microbiology lab. She introduced *E. coli* bacteria onto a series of floor tiles before placing gummy bears on the tiles for five seconds. In her experiment, the gummy bears did pick up the bacteria in just that short amount of time.

A more comprehensive picture was provided by a 2007 study published by Professor Paul Dawson of Clemson University in the *Journal of Applied Microbiology*. He contaminated several surfaces with salmonella – bacteria that is particularly associated with food poisoning – before dropping pieces of bologna sausage onto them. On a tiled floor, over ninety-nine per cent of the bacteria were

transferred to the meat within the first five seconds. However, on a carpeted floor, less than half a per cent of the bacteria ended up on the food. A wooden surface came out somewhere in between, with a transfer rate of between five and sixty-eight per cent. So, clearly the type of surface matters.

This research is corroborated by a 2014 study from Aston University in the UK. A team of biology students led by Professor Anthony Hilton dropped toast, pasta, cookies and sticky sweets onto carpets and both laminated and wooden floors, all of which were contaminated with *E. coli* and bacteria named *Staphylococcus aureus*. What Hilton's team found backs up Dawson's work – carpet is the surface least likely to transfer bacteria to dropped food. As a rule, dry foods also picked up fewer bacteria than moist foods. The researchers also found that the number of bacteria present on the food goes up by a factor of ten between three and thirty seconds after it has been dropped.

So, the bottom line seems to be this: the five-second rule is not entirely accurate, as bacteria begin to be transferred to food in the instant that it touches the floor. However, if you retrieve the food within five seconds, you have at least limited the damage. Certainly don't wait any longer. Really it is a judgement call – if you've dropped moist food onto a tiled floor, it is probably best to throw the food away. However, you might be tempted to invoke the rule if you've dropped dry food on carpet. One last thought: children and the elderly should probably dispense with the five second rule entirely, as their immune systems are generally weaker than that of the average adult.

HOW TO MAKE THE PERFECT CUP OF TEA

Such is our obsession with tea that it has played its part in several conflicts through history. See the Opium Wars between Britain and China, or the Boston Tea Party, which helped ignite the American Revolution in the eighteenth century.

Today tea is still big business. The global tea market in 2016 was worth an estimated thirty-eight billion dollars. It may surprise you to learn that Britain ranks only fifth in terms of the amount of tea consumed per person per year. It is beaten by Mauritania, Ireland, Morocco and Turkey – the Turks each get through 7.5kg (148lb) of the stuff every year.

With so much char being quaffed around the world, it is no surprise that the debate around how to prepare the best cup has caught the eye of the scientific community. In 2003, the Royal Society of Chemistry enlisted the help of chemical engineer Dr Andrew Stapley from Loughborough University in the UK in order to determine the best way to make a cuppa.

Dr Stapley's first tip is to dispense with the teabag and use only loose-leaf tea. Despite being used by the majority of us, the teabag is supposedly an accidental invention anyway. Tea merchant Thomas Sullivan was sending out tea leaves to his customers in little silk bags, but his customers mistakenly put the whole bag in hot water to make their drink. The convenient idea stuck and silk was quickly replaced with paper teabags.

According to Stapley, you should boil soft fresh water in a kettle whilst simultaneously heating a quarter of a cup of water inside a ceramic teapot in the microwave for one minute. This serves to warm the pot up and you should then pour away the teapot water once the kettle has boiled. Place one teaspoon of loose-leaf Assam tea per cup desired into the pot and immediately pour over the boiling water and stir. Dr Stapley recommends brewing for three minutes.

Now comes the real controversy. Stapley says to put chilled milk into a ceramic mug before you pour in the brewed tea. ISO 3103 – the tea edict published by the International Organization for Standardization – also calls for milk first. This is so that the milk is heated evenly by the incoming tea, preventing the proteins within the milk from clumping together and forming the skin you sometimes get on top of the tea. It also allows the milk to help cool the tea more quickly, as Stapley notes that the optimum drinking temperature is between 60 and 65°C (140 and 149°F) to avoid unsightly slurping. If necessary, he suggests putting a teaspoon into the drink for additional cooling power.

However, a word of warning: this advice applies to tea brewed in a pot. If you're using a teabag in the mug from the get-go, then whatever you do, don't put the milk in first with the teabag. The milk will lower the temperature of the water and the tea won't brew properly.

CHECKLIST

1. ☐ LOOSE-LEAF ASIAN TEA
2. ☐ SOFT WATER
3. ☐ FRESH, CHILLED MILK
4. ☐ WHITE SUGAR
5. ☐ KETTLE
6. ☐ CERAMIC TEA-POT
7. ☐ LARGE CERAMIC MUG
8. ☐ FINE MESH TEA STRAINER
9. ☐ TEA SPOON
10. ☐ MICROWAVE OVEN

1
2
3
4
5
6
7
8
9
10

TOP TEA TIPS ACCORDING TO SCIENCE:

- Use loose-leaf tea, not a teabag.
- Preheat a ceramic teapot in the microwave.
- Use one teaspoon of tea leaves per cup of tea.
- Brew for three minutes.
- Put chilled milk into the teacup ahead of the tea.
- Drink at a temperature of 60–65°C (140–149°F).

HOW TO ARRANGE THE FOOD IN YOUR FRIDGE

If you're anything like the average person, you probably don't give much thought to where everything goes in your fridge. Each time you return from the supermarket laden with heavy bags, you hurriedly place your items on random shelves and it is never the same arrangement as the last time you did it. But there's a problem with that approach: food waste.

SHELVES

1 COOKED MEAT

2 LEFTOVERS

3 MILK

4 YOGHURT

5 CHEESE

6 BUTTER

7 RAW MEAT

8 EGGS

9 FISH

FRIDGE DRAWERS

10 FRUIT

11 VEG/SALAD

FRIDGE DOOR

12 CONDIMENTS

13 JUICE

Estimates suggest that we waste thirty per cent of the food we buy, unnecessarily discarding over a billion tonnes every year. It isn't just the food either – think of the fertilizers and pesticides also wasted on food you don't eat, the greenhouse gases emitted by transporting it and the methane produced when it rots at the dump.

One way around this problem is to get the best out of your food and make it last as long as possible. You can do this by placing your shopping into the areas of the fridge with the optimum temperature for that particular foodstuff. That way you'll also cut down on your shopping bill, leading to more money in your pocket.

So what's the deal? Well, many of us are familiar with the phrase "hot air rises", but that's not strictly true. Warmer air doesn't move upward because it is hotter. Instead, cold air – which is heavier – sinks, forcing the warmer air out of the way. Nevertheless, this still means the bottom shelves of your fridge maintain a lower temperature than those at the top. So, you should store things like raw meat, fish and poultry in the bottom half of the fridge (ideally on the bottom shelf to prevent raw meat juices dripping onto other food). Conversely, the warmer top shelves are the place for things less sensitive to temperature such as soft drinks, cooked meats and any leftovers. Be sure not to put hot leftovers in the fridge, however, as these will quickly raise the overall ambient temperature.

But there is an area of the fridge that is even warmer than the top shelf: the doors. They get warmed whenever you open the fridge. So, even though many of us do it regularly, you really shouldn't be putting your milk in the fridge door – it will go off quicker. Put your milk and other diary products in the middle of the fridge and instead fill the doors with your condiments and less perishable items such as orange juice. Another common mistake people make (authors included!) is to store bread, fruit and wine on top of the fridge. This is a particularly warm area that will lead to food going off faster.

Don't forget that not everything belongs in the fridge either. A 2013 study published in the journal *Food Chemistry* looked at the aroma of tomatoes when they were kept at 4°C (39°F) and 20°C (68°F) respectively. They found that storing them at fridge temperature, even for short durations, had a detrimental effect on their aroma. So best keep them on the counter top. Potatoes, on the other hand, should be stored in a dark cupboard. At least that's according to researchers from the University of Idaho who found that the cool, dingy conditions best prevented the potato – which is still alive – from sprouting. Garlic and onions should also be placed in the cupboard. Eggs are a bit of a grey area. Whilst they are safe and last longer in the fridge, their porous shells means their quality can easily be affected by other smelly food in the vicinity.

One final no-no: don't put opened cans of food in the fridge. According to the University of Warwick, there is a higher chance of the metal being transferred to the food.

HOW BEST TO STACK THE DISHWASHER

Dishwashers are one of our most convenient timesaving inventions. For those of us lucky enough to have them, gone are the days of stooping over the kitchen sink furiously scrubbing dishes.

Dishwashers are better for the environment, too. A study by researchers at the University of Bonn in Germany found that dishwashers use half the energy, one-sixth of the water and less soap than the average person doing the same task by hand. Even the most frugal manual washer could only just keep up with the efficiency of the machine. Dishwashers do a better job of killing bacteria as well, because they can wash with hotter water than the human skin can handle. Plus, according to a study published in the *Journal of Applied Microbiology*, the average kitchen sponge is laced with 200,000 times more bacteria than a toilet seat and contains more faecal bacteria to boot.

So dishwashers certainly seem to be the way forward. But there are still ways you can geek up your dishwasher routine to get the best clean possible. Curious about the best way to stack objects inside the machine, engineer Raúl Pérez Mohedano and colleagues from the University of Birmingham, UK, filled a dishwasher with radioactive tracers to see how the water flowed inside. Their technique – positron emission particle tracking – is similar to the PET scans used by doctors to see how well parts of the body are working.

Pérez Mohedano's team results, published in the *Chemical Engineering Journal*, show that it is best to place plates in a circular arrangement, all facing inwards. This can be difficult, given the

conventional "row by row" setup found in most conventional dishwashers, but you can get a good approximation of this arrangement with a bit of jiggling around. It makes sense, when you consider that water is dispersed from an arm in the middle of the dishwasher that rotates in a circular fashion. Neatly marshalling your crockery in a rectangle, like

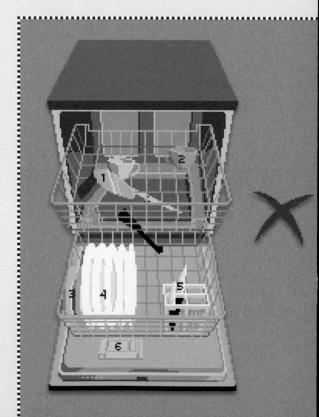

soldiers on parade, necessarily means that some will receive more water than others. In a similar vein, don't overfill the dishwasher, as this will prevent the water from moving freely around inside. The same applies to placing larger items in the path of the rotating arm.

Pérez Mohedano also recommends grouping your plates by type of stain. Protein stains such as egg yolk and custard need to be slowly hydrated in order to break down, but then don't require high water speeds to power them away. The best place for this type of cleaning was found to be at the sides of the bottom rack of the dishwasher. Carbohydrate stains such as potato and tomato need much more of a blast and should be placed where the water jet velocity is at its highest. Pérez Mohedano's findings suggest that this is in the middle of the top shelf, directly above the rotating arm. Alternatively, place the offending articles in a circle inside the ring of protein-stained plates on the bottom rack.

One final note on pre-rinsing your plates in the sink before putting them in the dishwasher: don't bother. The detergents have nothing to attach to and so will end up on your glasses and make them cloudy. You only need to scrape off any excess food and then you're good to go.

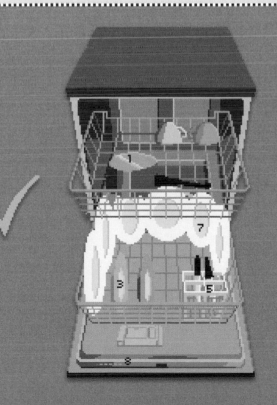

1 CARBOHYDRATE SOILED ITEMS IN THE MIDDLE OF THE TOP SHELF. HANDLES KEPT IN THE RACK

2 CUPS AND BOWLS PLACED FACE DOWN

3 SMALLER ITEMS PLACED IN THE MIDDLE

4 DON'T OVERFILL OR OVERLAP DISHES

5 PLACE KNIVES FACE DOWN IN BASKET

6 MAKE SURE DETERGENT FLAP ISN'T BLOCKED

7 ARRANGE PLATES IN A CIRCLE AROUND THE CENTRE

8 DON'T PRE-RINSE

HOW TO STOP CRYING OVER YOUR ONIONS

It is one of the most frustrating aspects of preparing a meal. You begin chopping onions, only for your eyes to fill up with so much water that you can barely see. You're left looking like you are grieving for the demise of a vegetable. In short, you're a blubbering mess.

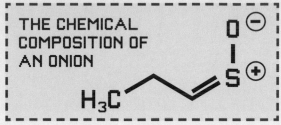

THE CHEMICAL
COMPOSITION OF
AN ONION

Back in 2002, a team of Japanese scientists cracked why slicing an onion has you reaching for the tissues. Publishing in the journal *Nature*, they described how the process starts with the sulphur the onion absorbs from the soil when it is growing underground. When you cut the onion, you break open some of its cells and release chemicals, which mix to form a substance with the trip-off-the-tongue name of "syn-propanethial-S-oxide". This is released into the air and comes into contact with your eyes, where it reacts with water to form an acid. This irritates your cornea, prompting your lacrimal glands to unload a deluge of tears to wash it away. According to biologists at Brandeis University in the United States, animals have been using similar defence mechanisms for 500 million years.

Given that it is such a ubiquitous problem, it is no surprise that there are a whole host of old wives' tales floating around about how to prevent the tears. They range from putting either bread or a silver spoon in your mouth, to burning a candle nearby. But, in reality, these endeavours are useless. What you actually want to be doing is preventing as much syn-propanethial-S-oxide reaching your eyeballs as possible. And there are several sensible ways that this can be done.

First up, make sure you cut your onion with a very sharp knife. This will limit the amount of damage you do to the onion's cellular structure, meaning fewer tear-inducing chemicals are released. More sulphur is contained in the root of the onion than the rest of the bulb, so keep this intact while you cut the rest. You can also slow down the rate at which those chemicals turn into syn-propanethial-S-oxide by lowering the onion's temperature – stick it in the fridge for a bit before you cut it. Some people suggest cutting onions underwater to help wash the chemicals away. While this would work, it sounds very hard to achieve in practice. Sharp knife plus slippery hands just doesn't seem like a good idea. You could, however, cut onions with a fan nearby or beneath an oven's extractor fan.

Of course, the only foolproof way of remaining completely tear-free is to stop any gas reaching your eyes at all. Although perhaps wearing a pair of swimming goggles while cutting onions is a step too far!

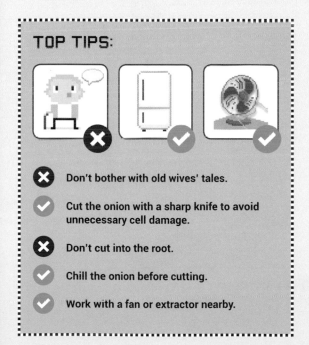

TOP TIPS:

✗ Don't bother with old wives' tales.

✓ Cut the onion with a sharp knife to avoid unnecessary cell damage.

✗ Don't cut into the root.

✓ Chill the onion before cutting.

✓ Work with a fan or extractor nearby.

HOW TO MAKE FOOD TASTE BETTER

We spend so much of our lives eating that we might as well make the most of it. Why waste time eating stuff that only tastes "all right"? With such a culinary gauntlet laid down, we've been searching high and low for unusual ways to make your food taste great.

Perhaps surprisingly, the environment around us when we are eating can have an effect on how our food tastes. This is the realm of Professor Charles Spence, an experimental psychologist from Oxford University who has collaborated with top chefs, including Heston Blumenthal. He calls his field "gastrophysics".

One of Professor Spence's key ingredients for a boost to the tastiness of your meal is music. Spence enlisted the help of seven hundred volunteers, asking them to rate various takeaway cuisines from one to ten whilst listening to six different musical genres. His conclusion was that certain types of music cause us to rate particular types of food more highly. Indie music, for example, boosted the ratings of

TOP TIPS:

 Pair up your takeaway with the right kind of background music (never Justin Bieber!).

 Eat from a round, white plate.

 Use heavier cutlery.

✔ **Delay eating by performing a quick task (e.g. taking a photograph).**

Indian food. If you don't ordinarily like Thai food or Japanese sushi, maybe you should try it again while listening to jazz music. Interestingly, the songs of Justin Bieber consistently lowered the enjoyment of almost every type of food studied (seriously!).

You'll also taste food differently according to how it is presented – particularly the type of plate it is served upon. Peter Stewart from the Memorial University of Newfoundland studied how food was rated for taste and quality when served on black, white, square or round plates. Round, white plates led to an increased rating for the same food. Even the cutlery you choose to eat with matters. In 2013, Spence published a study in the journal *Flavour* examining how the taste of food is affected by the

size, weight, shape and colour of the implements you eat it with. Want your dessert to taste sweeter? Use a small spoon. In a later study published in the same journal, and also involving Spence, participants were asked at the end of a three-course meal how much they would be willing to pay for what they had just eaten. Those eating with heavy cutlery offered fifteen per cent more than those using lighter knives, forks and spoons.

Our next piece of advice is controversial, one that might see you break one of the great social media taboos: photograph your food. We've all tutted at that friend who is always posting images of their breakfast on Facebook, but it might be making it taste better to them. A study published in the *Journal*

FOOD TASTES BETTER WHEN EATEN OFF ROUND, WHITE PLATES.

of *Consumer Marketing* found that photographing indulgent food such as red velvet cake prior to eating made it taste more delicious. Unfortunately, there was no such effect for healthy food. The researchers concluded that this taste top-up is due to stopping to savour your food before eating it. This backs up findings from a 2013 study in the journal *Psychological Science*, which found that delaying eating by performing a quick task boosted how participants perceived their meal. So, it seems it is the *taking* of the photograph that counts – there's no need to post it online as well!

PHOTOGRAPHING YOUR CAKE BEFORE YOU EAT IT INCREASES ITS DELICIOUSNESS.

TEMPERATURE AND TASTE

Lots of food packaging carries a familiar warning: "ensure food is piping hot before serving". Whilst that is to make sure the content is actually cooked, serving food that's still steaming might not be the way to go when it comes to maximizing flavour. At least that's according to a study led by Karel Talavera Pérez at the University of Leuven, Belgium and published in *Cellular and Molecular Life Sciences*. They found that taste perception decreases beyond 35°C (95°F).

And that wasn't the first time scientists discovered a link between temperature and taste. As far back as 2000, researchers Alberto Cruz and Barry Green found that as many as half the nerve cells responsible for taste respond to changes in temperature. What's more, their research – published in the journal *Nature* – found that heating or cooling the tongue can lead to the perception of taste even though the subjects weren't actually eating any food. It seems that approximately one in four of us are these so-called "thermal tasters". Warming the taste buds invoked a taste of sweetness; cooling the tongue left a sour taste.

Further work by researchers at Brock University in Canada has solidified this link between temperature and taste. Volunteers were asked to try chemicals designed to deliver hits of sweet, sour, bitter and astringent taste. These samples were either given at a temperature of 5°C (41°F) or 35°C (95°F). There was no difference in sweetness perception between the two, but astringency and sourness were both more

intense and lasted longer with the warm solution. Bitterness was more intense with the cold solution, but the flavour disappeared more quickly compared to the warmer alternative.

Given that temperature and taste are so intertwined, researchers led by Pauline Mony at the University of Arkansas wondered if the temperature of what we drink when we eat makes a difference. North Americans, for example, often have ice cold drinks with their food. In Europe, water is more likely to be consumed close to room temperature. Asians, on the other hand, will drink hot drinks such as tea alongside their meals. So the researchers fed participants dark chocolate after having washed their mouths out for five seconds with water that was at either a temperature of 4°C (39°F), 20°C (68°F) or 50°C (122°F). Their results, published in the journal *Food Quality and Preference*, indicate a decrease in the perceived intensities of sweetness, chocolate flavour and creaminess after consuming the coldest water. The researchers speculated that this may be one of the reasons why Americans show a preference for highly sweetened food – drinking cold water deadens the tongue to sweetness and so they have to put more in.

With a clear link between temperature and taste, you should experiment with how hot or cold you have your food and see whether it makes a difference for you. Certainly, it seems, piping hot might not always be best.

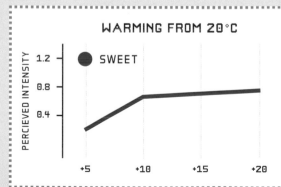

WARMING FROM 20°C

● SWEET

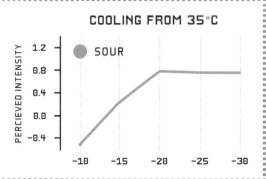

COOLING FROM 35°C

● SOUR

A BRIEF HISTORY

REAL GEEKS

CHARLES DARWIN
(1809–1882)

LEONARDO DA VINCI
(1452–1519)

ADA LOVELACE
(1815–1852)

ISAAC NEWTON
(1643–1727)

MARIE CURIE
(1867–1934)

1400   1500   1600   1700

FICTIONAL GEEKS

SHERLOCK HOLMES
(FIRST APPEARED 1887)

TONY STARK
(FIRST APPEARED 1963)

BRUCE WAYNE
(FIRST APPEARED 1939)

MR SPOCK
(FIRST APPEARED 1966)

THE DOCTOR
(FIRST APPEARED 1963)

OF THE GEEK

  ALBERT EINSTEIN
(1879–1955)

  PHILIP K. DICK
(1928–1982)

 RICHARD FEYNMAN
(1918–1988)

 STEPHEN HAWKING
(1942–)

 ROSALIND FRANKLIN
(1920–1958)

 STEVE JOBS
(1955–2011)

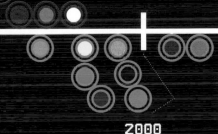

1800   1900   2000

  DR EMMETT BROWN
(FIRST APPEARED 1985)

  DANA SCULLY
(FIRST APPEARED 1993)

  LISA SIMPSON
(FIRST APPEARED 1987)

  DR SHELDON COOPER
(FIRST APPEARED 2007)

  REY
(FIRST APPEARED 2015)

CHAPTER 5

LEISURE AND SPORT

HOW TO ROCK AT ROCK, PAPER, SCISSORS

It's a time-honoured way of settling a dispute, but is rock, paper, scissors truly random? Science says it isn't, and you can use what researchers have found to your advantage.

You'd probably expect the three moves to crop up with equal probability over many games. However, a 1998 Japanese study found rock was played 35 per cent of time compared to 33 per cent for paper and 31 per cent for scissors. Thanks to a Facebook version of the game, scientists also have access to statistics from over 1.6 million online head-to-heads, including 10 million move choices. It backs up the Japanese study – 36 per cent rock, 34 per cent scissors, 30 per cent paper. People love to play rock.

So what should you play? That depends on who you are up against. Rookie players are most likely to open with rock, so you might try starting with paper. If your opponent knows this, however, they might try paper too and you'll draw. Worse still, they will commence with scissors and cut you in half. The advice of the pros seems to be open with

scissors. It certainly worked for Christie's auction house in London in 2005. A Japanese art collector could not decide whether to let Christie's, or their rival Sotheby's, sell his collection, so he asked them to settle it with a game of rock, paper, scissors. The daughters of a Christie's employee suggested going for scissors because everyone expects you to open with rock so they will start with paper. It worked. Sotheby's played paper, Christie's played scissors and walked away with millions of dollars' worth of commission.

Playing scissors might secure you the first round, but what if you are playing a match with many rounds? Keep things as close to random as possible. A seasoned opponent will jump on any patterns you start introducing into your play. The less your opponent can guess your next move the better.

TOP TIPS:

- Keep your moves random.
- Look out for your opponent playing a winning move again.
- When they lose, watch for your rival cycling through the moves in the order they appear in the name of the game.
- Play with your eyes closed.

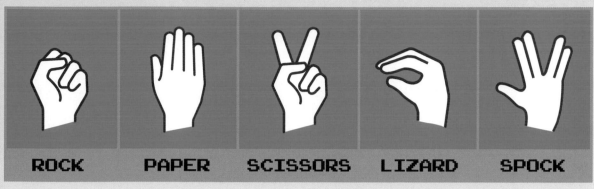

ROCK · PAPER · SCISSORS · LIZARD · SPOCK

It may seem like a game of chance, but it's really a game of psychology. A 2014 study, published in the journal *Nature*, analysed game-play at a Chinese tournament. They found that when somebody wins they often try the same successful move again – so if they win with rock they'll play it again. Conversely, when a player loses, they change their losing hand. What's more, their next choice followed the name of the game – if they lost with rock they switched to paper and then to scissors. Attempt to spot this pattern in your opponent's play and be careful not to fall foul of it yourself – keep things random.

One way to spice up the game is to introduce more possible moves. Inspired by the intuition that players who know each other well will end up cancelling each other out and draw over seventy-five per cent of the time, US software developer Sam Kass invented rock, paper, scissors, lizard, Spock – a game since immortalized in every geek's favourite comedy *The Big Bang Theory* (see diagram).

If all else fails, consider closing your eyes. A 2011 study found a lot more draws when one player could see and the other could not. However, on the occasions there was a winner it tended to be the unsighted player.

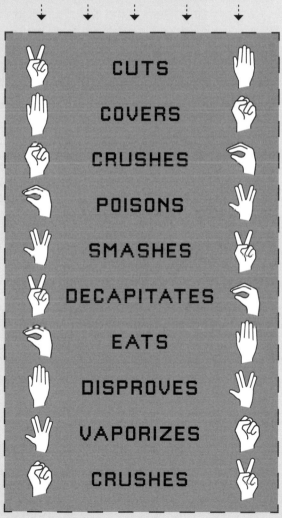

CUTS

COVERS

CRUSHES

POISONS

SMASHES

DECAPITATES

EATS

DISPROVES

VAPORIZES

CRUSHES

HOW TO KICK SOME SERIOUS ASS AT MONOPOLY

You know the drill. After all, Christmas Day is the same every year: too much turkey, another pair of socks and the realization that you've missed the Queen's speech while setting fire to a pudding. Then, just as you're sinking into a comfy chair, to partake in a bloated, mulled wine-induced doze, some distant relative perks up with, "Anyone fancy a game of Monopoly?" The question is: how do you win? Because nothing will blunt the enthusiasm of the person who suggested it like the bitter taste of defeat.

Let's start with a simple fact: the most occupied square on the board is Jail. This is obvious when you think about it - no other square has more ways to get to it. There's the "Go to Jail" square; a "Chance" or "Community Chest" card can send you to the clink; three double rolls and you're a convict; and even if you're just passing through ("Just Visiting"), you're still occupying the space.

So, now you know that your opponents are more likely to be coming from Jail than from anywhere else on the board, what evil schemes can you employ to catch them out? Which properties are they most likely to land on? Which are the best colours to buy? In short: buy orange. But here's why. Imagine your Great Uncle

TOP TIPS FOR WIPING THE BOARD WITH YOUR MONOPOLY OPPONENTS

- **THE ORANGE PROPERTIES** are the most likely landing spots for anyone taking a turn from Jail – the most occupied square.

- The single most frequented property is Trafalgar Square (red). Buy all the reds and DOUBLE UP YOUR RENT. Not as good as orange, but still a great money spinner.

- After Trafalgar Square, Kings Cross and Fenchurch Street stations are occupied more than any other properties. SNAP UP THE STATIONS.

- DON'T WASTE YOUR MONEY on Old Kent Road or Park Lane. They're the least frequented properties.

- The FOOTFALL ON THE UTILITIES (Water Works and Electric Company) is higher than most single properties. Definitely worth the investment.

- Once you have a set, BUILDING THREE HOUSES returns your outlay at a much higher rate than with two houses. So focus on getting three houses on each property in a set before developing other sets and you'll reap the rewards.

Bob's top hat is on the Jail square. When rolling two six-sided dice there is only one way he can make two: 1 and 1. However, there are six ways to roll a seven: 6 and 1, 1 and 6, 5 and 2, 2 and 5, 4 and 3 and 3 and 4. And thanks to the rule that rolling a double provides the player with another throw, there are actually even more ways to make seven. In fact, when you take doubles into account, the squares most likely to be occupied during a turn, in order, are seven, eight, six and nine spaces from Jail.

![Graphic showing 41% chance of landing on an orange property when starting from jail] 🏠 **41%**

CHANCE OF LANDING ON AN ORANGE PROPERTY WHEN STARTING FROM JAIL.

For those who like a good graph – and who doesn't? – here's the full picture:

A quick glance at a Monopoly board reveals that, from Jail, a move of seven squares will land you on "Community Chest". Fat use that is. But a move of six, eight or nine squares, the next most likely moves, will plonk you straight on each of the three orange properties. So while instinctively you might have plumped for the purple of Mayfair and Park Lane, you can make some easy money by buying up Bow Street, Vine Street and Marlborough Street. Anyone starting from Jail will have a forty-one per cent chance of landing on one of your properties by the end of their turn. Just don't tell Great Uncle Bob.

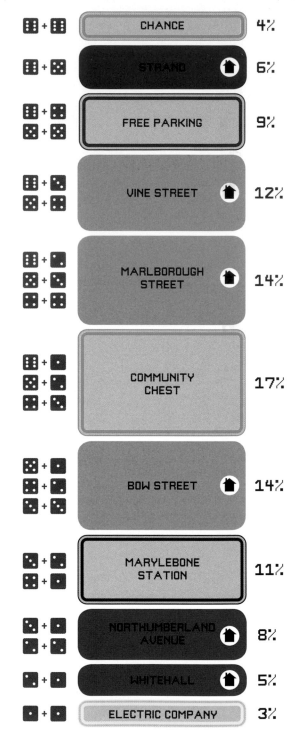

Dice	Square	%
⚃+⚃	CHANCE	4%
⚃+⚄	STRAND 🏠	6%
⚃+⚅ / ⚄+⚄	FREE PARKING	9%
⚃+⚅ / ⚄+⚄	VINE STREET 🏠	12%
⚂+⚅ / ⚄+⚄ / ⚃+⚅	MARLBOROUGH STREET 🏠	14%
⚃+⚀ / ⚄+⚁ / ⚃+⚂	COMMUNITY CHEST	17%
⚄+⚀ / ⚄+⚁ / ⚄+⚂	BOW STREET 🏠	14%
⚄+⚂ / ⚄+⚀	MARYLEBONE STATION	11%
⚄+⚀ / ⚄+⚁	NORTHUMBERLAND AVENUE 🏠	8%
⚄+⚀ / ⚁+⚁	WHITEHALL 🏠	5%
⚀+⚁	ELECTRIC COMPANY	3%

HOW TO WIN IN JUST FOUR TURNS

Monopoly is famous as a game that can rumble on for hours, but it is possible for it to be over incredibly quickly. Mathematicians have calculated that it might take as little as 21 seconds – or just four turns - for you to come out victorious. Although it does assume you can buy property on your first revolution of the board. Here's how such an ultimate game would play out:

PLAYER 1, TURN 1

Roll: Lands on: ELECTRIC COMPANY Action: None, but Doubles so rolls again

Roll: Lands on: TRAFALGAR SQUARE Action: None, but Doubles so rolls again

Roll: Lands on: COMMUNITY CHEST Action: Collects £200 (now has £1700)

which reads, "Bank error in your favour, Collect £200."

PLAYER 2, TURN 1

Roll: Lands on: INCOME TAX Action: Pay £200 (now has £1300), but Doubles so rolls again

Roll: Lands on: MARYLEBONE STATION Action: None

PLAYER 1, TURN 2

Roll: Lands on: PARK LANE Action: Buys it for £350 (now has £1350), but Doubles so rolls again

Roll: Lands on: MAYFAIR Action: Buys it for £400 (now has £950), but Doubles so rolls again

Roll: Lands on: WHITECHAPEL ROAD Action: Collects £200 for passing GO (now has £1150), then buys 3 houses for Mayfair and 2 for Park Lane at a cost of £1000 (now has £150)

PLAYER 2, TURN 2

Roll:  Lands on: CHANCE Action: Advances to Mayfair to pay rent of £1400 but only has £1300 so **goes bankrupt**

which reads, "Advance to Mayfair."

If you're thinking this sounds a little too good to be true then you're probably right.
This would only happen on average once every 253,899,891,671,040 games.

HOW TO STOP A TUNE PLAYING OVER AND OVER IN YOUR HEAD

If you've ever found yourself plagued by the music from some infuriating yoghurt advert or woken up humming the theme tune of *Jurassic Park*, then you've been struck by an earworm. That's the common name, invented by the Germans ("der Ohrwurm"), for a "musical itch" - a snippet of music that gets stuck in your head. Scientific studies suggest that this happens to around ninety per cent of us on at least a weekly basis.

There's no medical cure for an earworm. You can't go to your doctor and get them to put you on a twice-daily regimen of amoxicillin to clear up your ABBA infection. An earworm is a thought, often a product of mind-wandering, but it has no physical basis besides brainwaves. So the question is: how do you get rid of it?

Your mind is at its most susceptible to an earworm when it's off-guard, day-dreaming or failing hopelessly to concentrate on your latest influx of work emails.

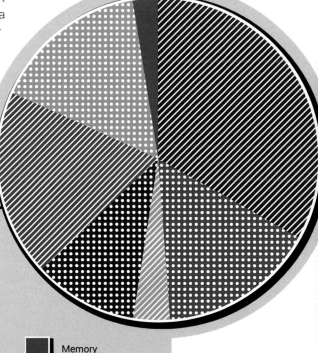

EARWORM TRIGGERS

 Listening to the song (duh)

 Association with person, sound or word

 Something that happened

 Thoughts or dreams

 Seriously, I have no idea

 Same old or default earworm

Memory

SOLUTION NO.1

Snap your mind back into action with something that takes up a bit more brain power. A series of evil experiments conducted by researchers at Western Washington University demonstrated this principle rather neatly. The scientists implanted Lady Gaga into the minds of willing volunteers by playing "Bad Romance" and "Paparazzi" at them and then attempted to extract their earworms by having the participants complete a range of puzzles. Five-letter anagrams worked best, though Sudoku puzzles were also successful when they weren't too difficult.

SOLUTION NO.2

The opposite strategy, recommended by no one with any genuine expertise, is to confront the earworm head on – listen to the tune that's been bugging the hell out of you. Well, you could try it anyway. We take absolutely no responsibility for the results.

SOLUTION NO.3

A third strategy, suggested by earworms expert Vicky Williamson from Goldsmith's University in London, is to dislodge the earworm with some other music. Play "Smooth Criminal". Put it on repeat until "Annie, are you okay?" is seeping out of your very pores.

Then there's the slightly baffling but nonetheless scientifically validated intervention...

SOLUTION NO.4

Chew some gum ... Huh? No, really. A 2015 study published in the *Journal of Experimental Psychology* found that chewing a stick of gum for three minutes often helped to relieve participants of insidious David Guetta and Maroon 5 infections.

There's one final option: pass it on to someone else. If you are now furiously trying to rid yourself of "Dancing Queen" or having trouble reading this last part due to the intrusion of "Annie, are you okay?" then congratulations, you have taken part in a successful exercise in earworm inception.

INSTANT EARWORM EXTRACTION KIT

- Puzzle book and pen.
- Previously tested displacement tune downloaded to your phone.
- Chewing gum.
- Access to the Internet to play the offending tune (only if desperate).
- An unsuspecting friend (for infection purposes).

HOW TO BE DEADLY AT DARTS

If you've ever played darts then you'll know it can be a very cruel game. The round board is divided into twenty scoring segments worth one to twenty points respectively, with a bulls-eye in the middle, which is worth fifty. A thin strip around the rim of the board is worth double points, and a similar strip approximately halfway out from the centre triples your score if you hit it.

Watch the pros on television constantly troubling the treble-twenty segment, and it looks like it wouldn't take much effort on your part to give them a run for their money. However, deadly accuracy is the key to their success, as the fiendish designers of the dart board have set up booby traps for the sloppy thrower, which are much like bunkers on a golf course.

A closer look and their wickedness becomes all too apparent. Yes, the twenty segment is where the points are

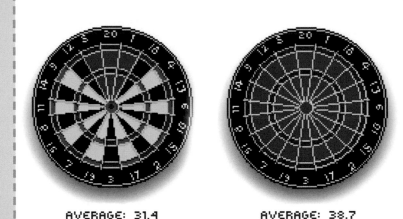

AVERAGE: 31.4   AVERAGE: 38.7

at, but only if you can hit it with consistency. Just a slight misjudgement will leave you languishing in the neighbouring segments, which are only worth five and one points respectively. Never fear, here at Geek Towers we've spent hours crunching the numbers and probing the probabilities to bring you the best tips for improving your dart score without hours and hours of practice.

In darts, each turn consists of three throws. Let's assume that on average you are accurate enough to get one dart somewhere in the number you are aiming for, but that the other two arrows stray into the adjoining ones. If your target is twenty, this means your score is likely to average twenty-six over many throws – if your arrows only land in the single scoring areas of those numbers. However, you'll end up in the doubles and triples from time to time, as well. Taking this factor into account, aiming for twenty will score you an average of 31.4 points per three darts.

Sounds pretty reasonable, right? But here's the thing: you can actually score more points by simply not aiming at all! We've worked out the average score you would get if you simply threw your three darts at the board at random: assuming all three land somewhere in any of the scoring areas, you're going to clock up an average of 38.7 points per trio of darts. So here's our first gem of information: if you're consistently scoring fewer than thirty-nine points per three darts by deliberately aiming, either practise more or trust in a bit of "hit and hope".

However, if you are good enough to get all three darts into a bank of three adjoining numbers, you

AVERAGE: 50.8

AVERAGE: 52.0

AVERAGE: 72.5

can beat this random score. You just shouldn't be aiming for twenty. Our analysis shows that you can maximize your score by aiming for seven. With the number sixteen and nineteen segments lurking either side, your lack of accuracy will actually be rewarded more often than not. Even if you just hit the singles areas, you'd be averaging forty-two points. If you take the odd double or triple into account, too, then your three-dart average jumps up to 50.8

points. Pretty healthy scoring, by any criteria.

More accurate players who are capable of getting two of their darts in their intended number and only one in an adjacent section should, unsurprisingly, be aiming for either the nineteen or twenty segments. And those who can get all three darts in one section should definitely return to aiming for the twenty. Although in that case, you probably don't need our help!

WHERE SHOULD YOU SIT AT THE MOVIES?

It's a common debate. You head along to your local cinema with a group of friends to catch the latest Hollywood blockbuster, but everyone has their own opinion about where they want to sit. Even with pre-booked seats, the same still applies when debating which row to book online. Luckily for you, we've been digging into the science behind cinema seating and can bring you the facts with which to win the next argument.

There are several ways to look at which is the best seat in a cinema. The most obvious criterion is a good view, so it is all about your sight line. Fortunately, the modern design of tiered cinema seating goes a long way to ensure that you can see the screen well from most seats in the theatre. However, if you are going to sit in the same spot for two hours or more then you definitely want to be comfortable. The Society of Motion Picture and Television Engineers (SMPTE) recommends the angle between you and the top of the screen should not be more than thirty-five degrees. So we've used the mathematics of trigonometry to put together a handy formula for you to work out the minimum distance you should be from screens of varying size:

> **MINIMUM DISTANCE = HEIGHT OF SCREEN / 0.7**

Standard screens range in width from 3 to 9 metres (10 to 30 feet). So you should sit a minimum of 4.3 metres (14 feet) back in the first case and 12.9 metres (42 feet) back in the second. IMAX screens can stretch to 15 metres (49 feet) tall. Here you'd need to be sitting at least 21.4 metres (70 feet) away.

However, it is possible to sit too far back. The SMPTE also suggests that the angle between you and the sides of the screen should be at least thirty-six degrees – the further back you sit, the narrower this angle becomes.

The formula this time therefore becomes:

> **MAXIMUM DISTANCE FROM SCREEN = WIDTH OF THE SCREEN / 0.65**

Typical movie screens range in width from 9 to 27.5 metres (29 to 90 feet). That means you should be no more than 13.8 metres (45 feet) away for the narrowest screens and 42.3 metres (139 feet) from the biggest.

If you don't want to take a tape measure along with you (who would!) then you can use your hands. Held at arm's length, your fist is ten degrees across. Stretch out your little finger and thumb on your other hand and that's twenty-five degrees. Put the two side by side and bingo, you've got a great way to judge thirty-five degrees!

The second factor to consider is sound. With spectacular modern sound technology, where should you sit in order to get the most vivid experience of

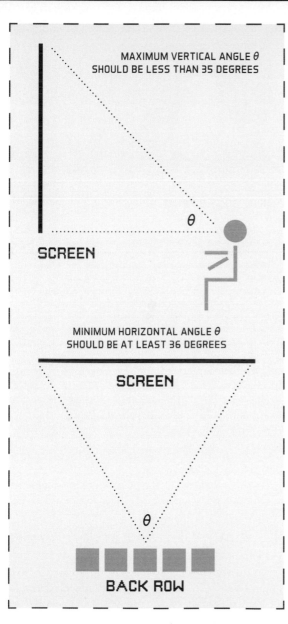

MAXIMUM VERTICAL ANGLE θ
SHOULD BE LESS THAN 35 DEGREES

θ

SCREEN

MINIMUM HORIZONTAL ANGLE θ
SHOULD BE AT LEAST 36 DEGREES

SCREEN

θ

BACK ROW

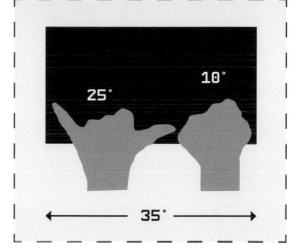

25°   10°

35°

KEY MOVIE-GOING TIPS:

✓ The angle between you and the top of the screen should not be more than thirty-five degrees.

✓ The angle between you and the sides of the screen should be at least thirty-six degrees.

✓ Either use our formulae to convert this into distance, or use your hands as a guide.

✓ Two-thirds of the way back in the centre of the cinema meets these criteria, with the added bonus of being in the audio "sweet spot".

✓ If these optimum seats are taken, then sitting on the left-hand side should give you more space in a show that isn't sold out.

THE POWER OF POPCORN

For many, a trip to the cinema just isn't the same without popcorn. In fact, it has become so much a part of the movie-going ritual that we'll continue to pile the stuff into our mouths even if it's stale. In 2011, University of Southern California researchers led by David Neal gave participants free tubs of popcorn as they entered the movie theatre. Some tubs contained freshly made popcorn, but in the others it had been left for a week to go stale. Afterwards, the researchers looked at how much popcorn had been consumed during the film. Their results, published in the *Personality and Social Psychology Bulletin*, show that regular moviegoers consumed the stale popcorn anyway. Those who didn't go to the cinema that often ate less of the stale stuff. We are, it seems, creatures of habit. We'll eat popcorn even if it's stale because we associate it with watching a movie.

That may sound gross, but a separate study found that eating popcorn might have an unexpected benefit: preventing advertisers from getting in your head. Researchers from the University of Cologne were behind a 2013 study in which ninety-six participants were invited to a movie screening that began with a series of adverts. Some of the participants were given popcorn to eat, whilst the others were given a swiftly dissolving sugar sweet. At the end of the film they were all tested to see if they exhibited any positive psychological responses to the adverts. No such responses were found in the popcorn eaters, but they were in the other group. According to the researchers, that is because when we encounter new information (such as a brand name) we subtly mouth it to ourselves to aid memory. As the popcorn eaters already had their mouths full, the adverts were less likely to influence them.

dramatic car chases, explosions and gunfire? The simple answer is to sit where the sound engineers sit when calibrating the speakers: two thirds of the way back and in the middle. You know that the sound here is guaranteed to be well balanced. Given that this position is also likely to satisfy the distance criteria for the best view, it seems fair to call that the best place to sit.

One last thing: what do you do if those prime seats are already taken when you turn up at the cinema or book online? We know that you want to be within those distance limits, but should you opt for the left or the right side of the theatre? According to a 2010 study by Matia Okubo, published in the journal *Applied Cognitive Psychology*, right-handed people have a natural preference for seats on the right-hand side. With at least seventy per cent of us thought to be right-handed, by opting for left-hand seats you're more likely to get peace and quiet from popcorn rustlers and canoodling couples.

WANT TO WIN AT SPORT? WEAR RED

Surely something as trivial as the colour you wear when playing sport can't have an effect on the result? Apparently it just might, according to a number of scientific studies on the subject.

Back in 2005, researchers Russell Hill and Robert Barton from the University of Durham, UK, analysed the results of the male combat sports in the 2004 Athens Olympic Games. They selected boxing, taekwondo, Greco-Roman wrestling and freestyle wrestling – the four sports in which the two competitors are randomly assigned red or blue protective equipment. Publishing their results in the journal *Nature*, they found that in fights in which the opponents were otherwise deemed to be equally matched, the red corner was victorious more often than not (*see graph*).

What is it about red that is swaying the matches in their favour? There are several theories. Hill and Barton suggest that wearing red may boost a competitor's testosterone levels, enabling them to perform better. Perhaps it is because their opponent is intimidated by the colour instead. A 2007 study published in the *Journal of Experimental Psychology* showed that even a fleeting exposure to the colour red impaired the performance of participants in written tests.

The third possible explanation does not involve either of the fighters, but the only other person involved: the referee. In a 2008 study, published in the journal *Psychological Science*, Norbert Hagemann showed a series of video clips of taekwondo bouts

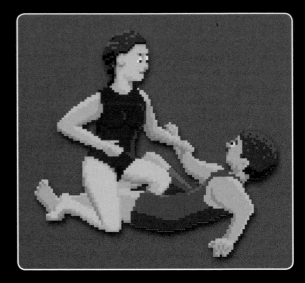

to forty-two referees and asked them to score the matches. They then showed the referees the same fights but in a different order, digitally doctoring the clips to swap red clothing for blue. The scores awarded by the referees were an average of thirteen per cent higher when the competitors were wearing red, even though it was exactly the same fight.

Clearly, colour makes a difference in these particular one-on-one disciplines, but what about team sports? Are they similarly affected? Barton and Hill set out to examine these as well in a later study published in the *Journal of Sports Sciences* in 2008. They analysed the results of soccer matches in England's top division between 1947 and 2003. According to their findings, teams wearing red won a disproportionate amount of home games compared to rivals donning other colours. The least successful colours were found to be yellow and orange, with blue and white somewhere in between. However, shirt colour had no effect on a team's performance away from home. It is worth saying that a similar exercise by German economists analysing one season of the German Bundesliga found no such link.

However, it isn't just soccer in which the effects of the colour red crop up. A 2012 study in the journal *Sport in Society* analysed thirty years of Australian rugby league matches. Although the authors warn that more work needs to be done, they did find a strong positive relationship between winning and the difference in redness between the kits of the home and away teams.

It seems that for whatever reason there is something in the idea that wearing red helps you to win, particularly in one-on-one combat sports. So, while you shouldn't rely solely on the colour of your kit for victory, you could do a lot worse than pull on a red jersey.

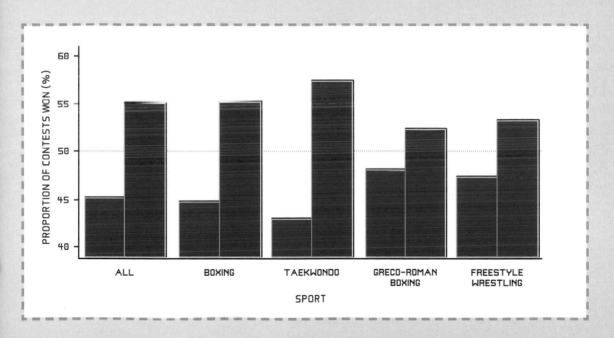

HOW TO BECOME A MASTER OF STONE SKIMMING

The pastime of stone skimming is ancient, with the Romans having been known to amuse themselves by skimming seashells. The idea is simple: pick up a stone lying on a beach or around a lake and launch it across the water so that it skips off the surface as many times as possible.

But what makes for a good skimming technique? Back in 2003, a team of French scientists led by Christophe Clanet endeavoured to find out. To simulate stone skimming in a controlled way, they fired flat aluminium discs into a pool of water two metres (6.5 feet) long. They varied the angle of impact, the speed of the discs and their rotation speed. Their findings, published in the journal *Nature*,

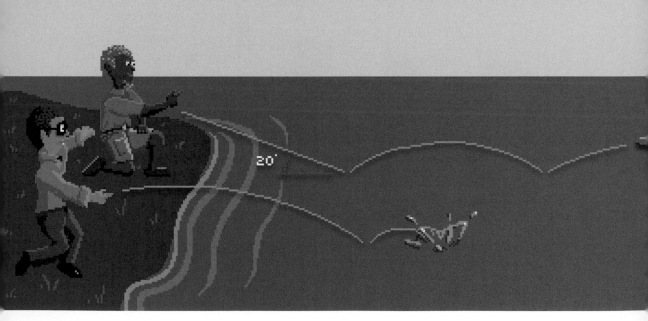

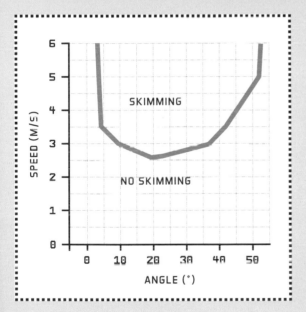

The reason for this finding was probed in an earlier study published by one of Clanet's co-authors – Lydéric Bocquet – in the *American Journal of Physics*. He found that a skimmed stone spends one hundred times more time in the air than it does in the water, but that every time it bounces it sinks a little deeper and the drag it experiences from the water increases. Eventually the stone doesn't have enough energy to burst back out of the water. Clanet's team found that rotating discs were more stable in flight and more likely to resist falling into the water.

However, it is the optimal angle and speed that are likely to be less familiar and yet crucial to success. The magic angle is twenty degrees between the stone and the water. To maximize the number of bounces at this angle, a stone should be launched at a speed of at least 2.5 metres per second (5.6 miles per hour, *see graph*).

The world record for the number of consecutive skips made by a stone is currently held by Kurt Steiner, who racked up a staggering total of eighty-eight bounces in September 2013 (watch the YouTube video; it's insane!). He gives the following advice when it comes to selecting your stone. It doesn't have to be perfectly round, but it should be smooth and have a flat bottom. It should weigh around one hundred to two hundred grams (three and a half to seven ounces) and be sixty to sixty-five millimetres (2.3–2.5 inches) thick. A previous record holder – Jerdone Coleman McGhee – cautions against launching your stone too far into the water before the first bounce, suggesting it should splash down within four and a half metres (fifteen feet) of the throwing point.

If you take these tips on board and feel confident in your new-found skimming skills, then you might consider taking on the rest of the world at the annual World Stone Skimming Championship held near the Scottish town of Oban. However, they rank the hundreds of competitors by total distance travelled, not the number of skips it takes to get there...!

can help you become a better stone skimmer.

Their first two pieces of advice won't come as a surprise to seasoned throwers: flat, round stones work the best. This is because they maximize the surface area in contact with the water for each bounce. Try and find stones roughly five centimetres (two inches) in diameter. You should then spin the stone with your fingers as it leaves your hand.

HOW TO BUILD THE PERFECT PAPER PLANE

It is a much-loved childhood pastime, and an activity perfected by many a bored student sitting at the back of a stuffy lecture hall. The attraction of paper planes is clear - you couldn't need simpler materials. Just fold a piece of paper in the right way and you can launch your creation far into the distance.

For one man, these flights of fancy have become an obsession. The world record for the furthest paper plane journey – some 69.14 metres (227 feet) – is currently held jointly by John Collins (who designed it) and Joe Ayoob (who threw it). Collins' design is lovingly named "Suzanne", and he has shared the secrets behind his craft in order to inspire others to take up the activity. You can view videos of both the record-breaking flight and him building the 'plane on his YouTube channel (*ThePaperAirplaneGuy*).

Let's start with material – Collins uses 100 gsm A4 laser paper. He also recommends that you fold the paper on a surface that is at least as smooth as the paper – he uses glass. Such is his attention to detail, that Collins carefully selects the perfect page by holding it up in front of a light bulb to look for any defects that might alter the plane's aerodynamics. If the heat from the light bulb starts to warp the paper slightly, then always make sure you fold against this curvature. Also, ensure these folds are as crisp as possible, by using a ruler or a folding bone to help you.

Once you've finished folding (*see diagram*) you should clip the plane in place and then use fourteen very small pieces of tape (2–4mm) to join the layers together to aid with stability. Upon completion of your flying machine, particular parts of it should have very specific angles. The angle at which the wings meet the tail from above should be 165 degrees, as should the angle where the wings meet the nose. However, the angle at which the wings join the main body of the plane (in line with the wing tips) should be 155 degrees.

Folding your plane by following these rules should give you an aircraft to rival any in the world for distance. Although we have, of course, only been talking about throwing it the furthest. There is another metric by which you might judge the best plane: hang time. That record is held by Japanese origami enthusiast Takuo Toda, who in December 2010 managed to keep his creation aloft for an amazing 29.2 seconds.

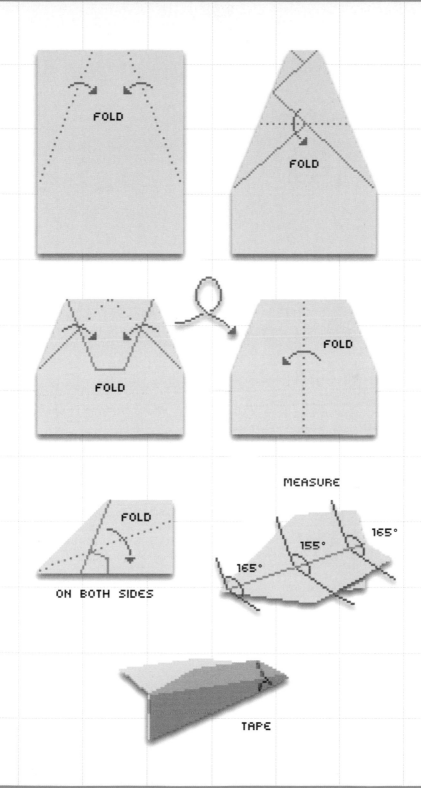

FOLD

FOLD

FOLD

FOLD

FOLD

ON BOTH SIDES

MEASURE

165°

155°

165°

TAPE

CHAPTER 6

TRAVEL

HOW TO NAVIGATE WITHOUT GPS

These days we rely on technology for almost everything. It means some everyday items that were once commonplace now sit on shelves gathering dust. Take the good old A-Z city street map. It used to be the go-to way of navigating the busy streets of a sprawling urban metropolis. However, the spread of GPS-enabled smartphones has seen sales of paper maps plummet. Satnav systems in cars have largely replaced the road atlas, too.

But these high-tech systems are not infallible. One giant solar storm, for example, could cripple the fleet of GPS satellites that we rely on to help us find our way around. In a more apocalyptic scenario, an asteroid strike or a widespread pandemic could send us right back to the Dark Ages. Maybe you just run out of battery or someone stole your phone. Could you find your way around without the aid of modern technology? Fortunately, the geeks of yesteryear worked out ways to navigate by the Sun and the stars. Here we share that knowledge with you – you know, just in case...

BY THE SUN

For centuries we believed that the Earth was the centre of the solar system and that the Sun, planets and stars bent to our will. After all, we don't feel the Earth spinning and it is the Sun that appears to move across the sky. Today, of course, we know that it is in fact the other way around. Nevertheless, the Sun's apparent voyage across the sky is an invaluable tool for navigation when all modern methods fail.

Due to the fact that the Earth rotates on its axis, the Sun always appears to rise towards the east and set towards the west, no matter which hemisphere you are in. However, the hemispheres differ in where they see the Sun go in between rising and setting. Those above the equator see it rise to its highest point due south, whereas those in the southern hemisphere will see it reach the top of its climb due north. Either way, the shadows created as the Sun travels across the sky appear to move in the opposite direction – a fact exploited by ancient civilizations to build sundials in the centuries before mechanical clocks.

However, the Sun only rises due east and sets due west at the equinoxes in March and September, so using these shadows acts as a much more reliable compass. First, you'll need a stick about a metre (one yard) long, which you stake into the ground. Then, mark the tip of the shadow it creates with a stone. Wait for the Sun and the shadow to move slightly (give it at least fifteen minutes). Place a second stone at the tip of the new shadow. Then, with your back to the Sun, place your feet against the stones. In the northern hemisphere you'll be facing due north and in the southern hemisphere due south. It is then easy to work out the other compass points from there.

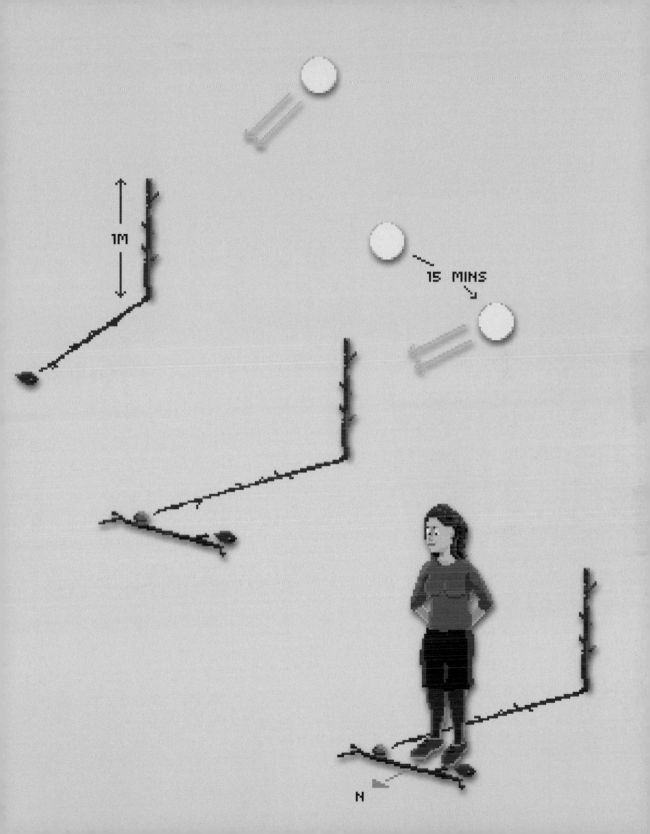

1M

15 MINS

N

NAVIGATING IN THE NORTHERN HEMISPHERE
FIND "POLARIS"

"POLARIS"

NORTH
POLE

"THE PLOUGH"

SOUTH
POLE

"THE SOUTHERN CROSS"

NAVIGATING IN THE SOUTHERN HEMISPHERE
USE "THE SOUTHERN CROSS"

BY THE STARS

Obviously, the shadow method only works during daylight hours. Fortunately, however, the stars can still help you find your way at night. Those in the northern hemisphere are lucky enough to have a pole star – named Polaris. As the Earth rotates, our view of space changes and so it seems as though the stars are moving across the sky. Polaris, however, sits almost directly above our axis of rotation and so to us barely seems to move. This steadfast nature means that it is a permanent beacon to the North Pole (it does move over thousands of years, but for us short-lived humans it might as well be permanently stationary). Just draw a line straight down to the ground to locate the northern horizon.

People often mistakenly say that Polaris is the brightest star in the night sky, but it is not even close to being so. Perhaps they are mixing up importance with brightness. Anyway, because Polaris isn't actually that bright, it pays to have an easy way to find it. Such a method is provided by a set of seven nearby stars in the shape of a giant saucepan, known variously around the world by names that include the Plough or the Big Dipper.

If you take the two stars on the opposite edge to the handle of the pan, and extend the line between them upwards, you are directed straight to Polaris. This star has the added navigational bonus that it can tell you where you are in the northern hemisphere. If you were standing on the North Pole, then it would be directly overhead. Stand on the equator instead and it would languish on the horizon. So, you can estimate how many degrees you are above the equator by measuring the angle between the ground and Polaris.

Unfortunately, southern hemisphere observers cannot see Polaris and have no pole star of their own. Instead, you should find the constellation of Crux – The Southern Cross. Take the longest part of the cross and extend that line beyond the bottom of the constellation. Continue for four and a half cross lengths and you'll arrive at the position of the south celestial pole. A line drawn to the ground will indicate due south.

HOW TO MAKE A SUNDIAL

In the days before clocks and watches, people could track the time by monitoring the shadows cast by the Sun as it appears to move across the sky due to the rotation of the Earth. The shadows are, in effect, the hour hand of the clock.

To make your own sundial you will need: card, a pen/pencil, scissors, glue, a compass and a ruler.

First create the gnomon – the triangular bit that sticks up. The key thing is that the angle at the base of the triangle must match your latitude – how far you are (in degrees) from the equator (see A). Adding a small rectangular flap on the bottom will allow you to glue it to the baseplate. Make sure the right angle is in the middle of the baseplate.

Write the number 12 at the top and use your compass to find south (or north in the southern hemisphere). Point the bottom of your sundial that way at noon and the shadow of the gnomon will fall on the number 12 (see B). Wait for the shadow to move and mark its position on the other hours. Draw in the lines with the ruler.

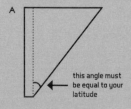

this angle must be equal to your latitude

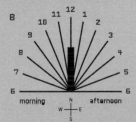

WHICH COLOUR CAR SHOULD YOU BUY?

Surely a car is just a car, right? You'd think that the colour you plump for is only a matter of taste. However, studies have shown that there are some potential benefits and downsides to particular hues.

According to research published in the *British Medical Journal*, silver cars are much less likely to be involved in accidents than vehicles of other colours. The study investigated road traffic accidents in and around Auckland, New Zealand, and found that of the most popular colours, silver cars were involved in the fewest incidents. They were fifty per cent less likely to have accidents compared to white vehicles. Yellow, grey, red and blue cars carried the same risk as white, with black, brown and green cars involved in the most accidents. This colour association remained even when the researchers accounted for other major variables. Lead researcher Sue Furness speculated that it might be down to the increased reflectivity of silver cars.

A later, more comprehensive study was undertaken by researchers at Monash University in Australia. They looked at a total of 850,000 traffic accidents over a twenty-year period. When it comes to the safest colours, they disagreed with the New Zealand study, citing yellow, white and gold cars as those involved in the fewest accidents. They did, however, back up the idea that black cars aren't the way to go, finding that during the day they were twelve per cent more likely to be caught up in a crash than white cars. At dawn or dusk this figure rose to forty-seven per cent.

Black cars also come out badly when you consider how attractive your car is to thieves. Economist Ben Vollaard of the University of Tilburg looked at the vehicles stolen in the Netherlands between 2004

RELATIVE RISK OF TRAFFIC ACCIDENTS OCCURING AT DAWN/ DUSK FOR DIFFERENT CAR COLOURS COMPARED TO WHITE

and 2008. His analysis revealed that black, blue and silver/grey cars were most likely to be pinched. Given that these are also some of the most popular car colours worldwide, it seems that thieves are stealing what they know they can sell on.

To make matters worse, a 2011 study published in the journal *Applied Energy* suggests that dark-coloured cars are less environmentally friendly than their lighter counterparts. The researchers from Lawrence Berkeley National Laboratory in California parked two versions of the same car – one black and one silver – facing directly towards the sun. After an hour they measured the vehicles' internal temperature, finding that the black car was 6°C (11°F) hotter. That means black car drivers would need use to more air conditioning to counteract the additional temperature boost. Further analysis by the team found that, due to lower air-conditioning usage, the lighter-coloured car would produce two per cent fewer carbon dioxide emissions.

So, whilst there doesn't appear to be a definitive colour of car that you should buy, it seems as though you might consider avoiding black – because they are involved in more accidents, more thefts and create more pollution.

0.66 PINK
0.82 MAROON
0.88 YELLOW
0.96 BLUE
0.99 CREAM
1.01 PURPLE
1.02 RED
1.03 GREEN
1.04 GOLD
1.12 BROWN
1.15 SILVER
1.21 ORANGE
1.25 GREY
1.47 BLACK

White = 1.00

HOW TO MANAGE JET LAG

Living in the twenty-first century certainly has its perks. Gone are the days when you could live out your entire life and never leave your home country. International travel has never been so accessible, but it does come with a major drawback: jet lag.

The condition – which leaves sufferers feeling groggy, nauseated and disorientated – is caused by the fact that jumping between time zones plays havoc with our body clocks. This internal biological time-keeping mechanism – known as our circadian rhythm – is governed by how much light we are exposed to and regulates many of our body's important processes. Jet setting knocks this cycle out of whack, with nasty consequences. Travelling eastwards hits you particularly hard, because the day is shortened and the body finds it even harder to adjust.

So, what can you do about it? The bad news is that there is still no surefire way to prevent jet lag entirely. However, there are things you can do to help your body settle back into a synchronized circadian rhythm more quickly. For example, abstaining from alcohol and coffee is said to prevent further disruption to your sleep/wake cycle.

According to the majority of research in this area, the biggest difference you can make is by starting to adjust your body clock towards your destination time zone in the days before you leave. However, this must be done gradually, as studies – including one in *The Journal of Clinical Endocrinology & Metabolism* – have shown that you can only move your clock by one hour per day if you're travelling east and by two hours a day if travelling west.

It all comes down to melatonin – the sleep hormone – which your body produces in response to a decreased amount of light entering your eyes. So, in order to adjust your sleep/wake cycle, you will need to seek out or avoid light at certain times of day. One way to do this is to shift the times at which you go to bed or wake up. Your usual sleeping pattern is important here, because what you really want to synchronize with your destination is the time at which your body temperature is at a minimum, as this is when you are sleepiest. This is normally three hours before you wake up if you sleep for more than seven hours, and two hours before if you sleep less than this.

There are websites and apps that can help you create a personalized plan to help you adjust your sleep routine, based on both your normal sleeping pattern and your start and end destinations. We have worked up the typical schedule for a flight from London to Bangkok (a time difference of +7 hours). We've assumed that the flight leaves at midday on a Thursday (London time) and arrives at 7am (Bangkok time) the next day. We've also assumed that you usually sleep from eleven o'clock at night to seven o'clock in the morning. Note: if you are flying west rather than east, you reverse the process and go to bed progressively later before departure.

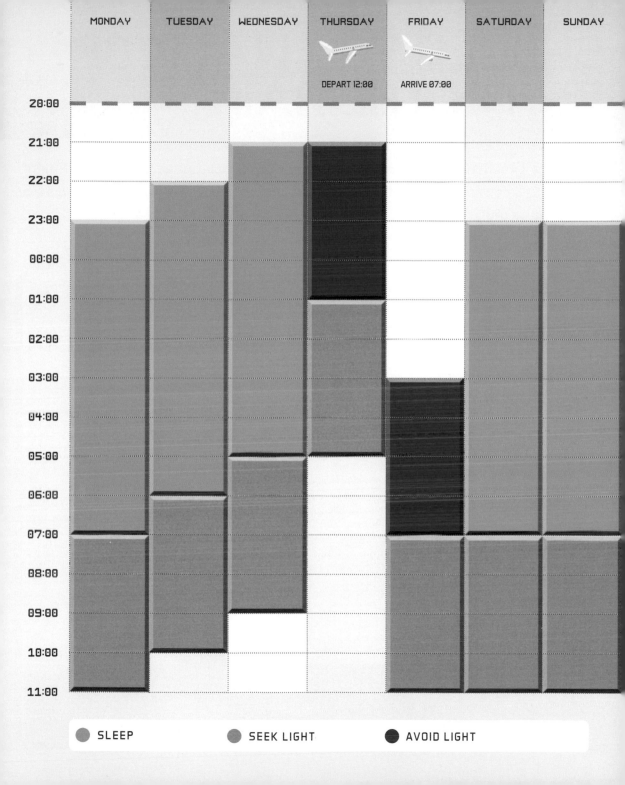

SHOULD YOU WALK OR RUN IN THE RAIN?

This is a question that you are sure to have contemplated at some point. You're caught in a downpour without a coat or an umbrella: what can you do to make sure you stay as dry as possible?

On the face of it, you might imagine that running would be the best thing to do, simply because that way you'll get out of the rain quicker. However, whilst running means you will be hit by fewer raindrops from above, more raindrops from the side will still hit you as you run into them.

So, which really is best? Thankfully, there is a mathematical formula to help us out:

$$\text{OVERALL WETNESS} =$$

$$(\text{WETNESS PER SECOND} \times \text{TIME IN THE RAIN})$$

$$+$$

$$(\text{WETNESS PER METRE} \times \text{DISTANCE TRAVELLED})$$

Let's say, then, that you want to get to the nearest sheltered doorway fifty metres (fifty-five yards) away. Whether you run or walk, you can't shorten that distance. So, the wetness contribution from the second half of the equation is fixed. Your only remaining option is to minimize the amount of time that you spend in the rain, and so the faster you get to the doorway the better – RUN!

Franco Bocci is a physicist who has extensively investigated this conundrum. In 2012, he published a paper in the *European Journal of Physics* that, in general, backs up this idea. However, he also found that it isn't quite as black and white as it first appears. He argues that if the rain is falling directly downwards, or the wind is blowing the rain into your face, then running so that you spend as little time in the deluge as possible is still the way to go. However, if the rain is blowing into your back, your optimum strategy is to run at the same speed as the wind – even if this means spending more time in the rain than running at full pelt. Bocci also looked at whether body shape plays a role, particularly when the wind is blowing from the side. His analysis shows that larger people should run at full speed, but thinner people should return to matching the speed of the wind.

Of course, the other factor to consider is running on a wet surface. Should you slip over, you're not only likely to get wetter than if you'd walked, but you could hurt yourself, too. We'll leave you to weigh up the options...

TOP TIPS:

- If the rain is hitting you from above or from the front, run as quickly as possible.

- If the rain is hitting your back, run at a speed closely matching the wind speed.

- Larger people should run as quickly as possible if the rain is coming in from the side.

- Thinner people should match the speed of a sideways wind.

HOW TO LEARN ANOTHER LANGUAGE

When travelling, speaking the local language can often make all the difference. However, according to 2013 research by the British Council, three quarters of Brits can't hold a conversation in any other language than English. A 2001 Gallup poll of Americans came to a similar conclusion.

So, how can you improve your chances of picking up a language that isn't your mother tongue? It seems that simply listening to native speakers is a good start. Back in 2010, scientists from the University of Cambridge used electrodes to study the brain activity of sixteen volunteers. First, researchers looked at how participants' brains responded when they heard a common word in their native language. Then they switched to playing them a fake word – one the researchers had invented. Participants heard this new word one hundred and sixty times in fourteen minutes. Publishing their results in the *Journal of Neuroscience*, the scientists found that by the end of this process the participants' brain activity in response to the fake word matched that of the common word. In less than a quarter of an hour, a new network of neurons had been created to remember this new word. Repetition, then, is key.

In that study, participants didn't even have to speak the word. But when you do venture into saying your new phrases out loud, it is important to get feedback. A study published in the journal *Cognitive, Affective and Behavioral Neuroscience* revolved around teaching Japanese speakers the difference between "r" and "l" in English – something they find particularly difficult. The volunteers were played recordings of "lock" and "rock" in a random order.

They were asked to press "L" or "R" on a computer keyboard, depending on what the volunteers thought they had heard, but they only got it right slightly more than half the time. But then the researchers started to introduce feedback – a green tick or a red cross on the screen. Within an hour the participants' accuracy hit eighty per cent.

However, it seems that nothing can beat immersing yourself in another language. Researchers at the Cognition of Second Language Acquisition Laboratory, part of the University of Illinois, conducted an experiment with two different groups of participants. Both groups were taught a new language that the researchers had invented. One group learned by studying the rules of the language, with the other group picking it up by being immersed in it, much like when we acquire our primary language as children. The brain activity of the second group was more like that of native speakers. The researchers then tested the participants again six months later. Despite the fact that they could not have had any further exposure to the language – it was made up, after all – the immersive learners performed better than the other group.

If you are successful in learning a second language, the pay-off could be greater than a better travelling experience. Research has shown that it can have a boosting effect on your brain. A 2014 study published in the *Annals of Neurology*, looked at 853 participants over many decades. Their cognitive abilities were first tested in 1947, when they were just eleven years old. Researchers caught up with them between 2008 and 2010 and tested them again. Those that could speak a second language did significantly better than expected, even if that additional skill had been acquired in adulthood.

TOP TIPS:

- Listen to new words repetitively over a short period of time.

- Get feedback on your pronunciation to learn faster.

- Immerse yourself in the conversation of native speakers.

WHEN TO BUY AN AIRLINE TICKET (AND WHERE TO SIT ON BOARD)

Nobody likes to pay more for something than they have to, and holidays are no exception. The way airline tickets are priced often seems akin to magic, fluctuating day-to-day - or even hour-by-hour - in a seemingly indiscernible way. Fortunately, thanks to the Internet, there is a wealth of data to analyse in order to find the best way to save money. After all, the less you spend on flights, the more margaritas you can drink on the beach when you arrive.

In January 2013, the website cheapair.com looked at more than half a billion online search records for domestic flights in the US for the previous year. Their analysis tracked the price of a ticket for the thirty weeks before the flight. On average, the fare was cheapest seven weeks before take-off (*see graph*). Analysis of international flights pegged the optimum date as eighty-one days prior to departure. Bear in mind that this is an average and the best time for individual flights will vary. It is, however, a good rule of thumb. It is also backed up by a similar analysis from momondo.co.uk. They found that the cheapest time to buy flights leaving the UK was fifty-three days before, cementing the seven-week rule. SkyScanner also recommends seven weeks.

With all this data available, it is possible to look at other questions, too. For example, is there a particular day of the week on which you should book flights in order to receive the greatest saving? The answer is a bit murky. For years, there has been a bit of an urban myth doing the rounds that you should book your tickets on a Tuesday. According to the

Expedia 2015 Air Travel Trends report, Tuesday is the cheapest day to book, but it doesn't really make that much difference. On average it is about five per cent cheaper to book on a Tuesday compared to the weekend (*see table on page 138*). However, the 2016 version of the same report threw a spanner in the works – Saturday and Sunday had overtaken Tuesday as the best days to book (although Tuesday was still the cheapest weekday). Friday was the most expensive day in the 2016 report and the most expensive weekday in the 2015 report, meaning our best advice is to avoid booking on a Friday.

What about which day you actually take to the skies? The 2015 report found that, for long-haul flights, return tickets were cheapest when flying out on a Thursday and returning on a Monday (an average of $798). Compare this to the most expensive option – an average of $999 for heading off on a Friday and coming back on a Saturday – and the difference is twenty per cent. For short-haul flights, leaving on a Saturday and returning on a Tuesday was consistently the cheapest option (an

	MON	TUE	WED	THU	FRI	SAT	SUN
FEBRUARY	1	2	3	4	5	6	7
	8	9	10	11	12	13	14
	15	16	17	18	19	20	21
	22	23	24	25	26	27	28
MARCH	1	2	3	4	5	6	7
	8	9	10	11	12	13	14
	15	16	17	18	19	20	21
	22	23	24	25	26	27	HOLIDAY

AVERAGE TICKET PRICE BY ADVANCE PURCHASE		
DAY OF WEEK	<21 DAYS	21+ DAYS
SUNDAY	$549	$539
MONDAY	$562	$520
TUESDAY	$561	$515
WEDNESDAY	$564	$517
THURSDAY	$568	$518
FRIDAY	$571	$522
SATURDAY	$561	$543

average of $369). Departing on a Sunday and flying back on a Monday would hit your wallet the hardest (an average of $493).

So, you have bought your ticket, and hopefully grabbed yourself a bargain. Next comes the decision about where to sit on the plane. Obviously, the optimum seat is going to vary from person to person, based on what a good seat means to you, but safety is probably high on many a person's list. Now, we should stress that flying is incredibly safe. In fact, according to Ian Savage, an economist at Northwestern University, it is the least dangerous way to travel. He found that flying causes just 0.07 deaths per billion miles travelled by passengers (*see below*). That compares to 0.43 for trains, 7.28 for cars and 212.57 for motorcycles. So, it is very unlikely that you'll be involved in a plane crash. That said, if you are concerned about safety, then you might want to sit at the back of the plane.

In 2007, the website popularmechanics.com looked back at the records for plane crashes in the US since 1971 – a total of twenty incidents. They compared fatalities and survivors to where they were sitting on the plane. In eleven out of the twenty accidents, those sitting at the rear were most likely to survive. On only five occasions did more people at the front survive (the rest showed no difference, or it could not be accurately determined). In 2015, *Time*

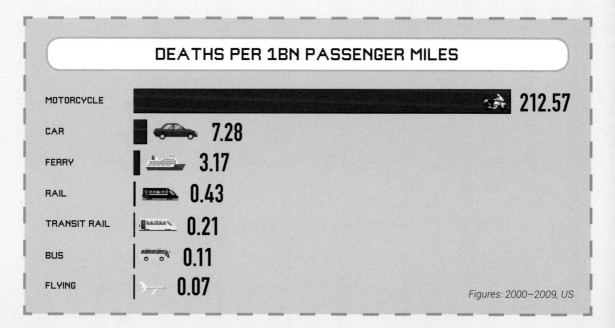

DEATHS PER 1BN PASSENGER MILES

MOTORCYCLE	212.57
CAR	7.28
FERRY	3.17
RAIL	0.43
TRANSIT RAIL	0.21
BUS	0.11
FLYING	0.07

Figures: 2000–2009, US

magazine undertook a similar exercise, analysing the seating charts of seventeen crashes that occurred between 1985 and 2000. They found that the rear third of the plane had the lowest fatality rate, at thirty-two per cent. That compares with thirty-nine per cent in the middle third and thirty-eight per cent at the front.

You should also sit in an aisle seat within five rows of an exit. That's according to a 2011 study by Ed Galea from the University of Greenwich in London. After looking at data from over one hundred crashes, he concluded that passengers in those seats are more likely to survive. He says that most deaths aren't caused by the crash itself, but because people cannot escape from the wreckage quickly enough. So, should the unthinkable happen, these seats ensure that you will be able to make

a swift exit. Combining all this information, we suggest sitting in an aisle seat, at the back of the plane and within five rows of an exit – you know, just in case...

TOP TIPS:

✓ **SHORT-HAUL FLIGHTS:** buy your tickets around seven weeks in advance and fly out on a Saturday and return on a Tuesday.

✓ **LONG-HAUL FLIGHTS:** book earlier than seven weeks ahead and fly out on a Thursday and return on a Monday.

✗ Don't book on a Friday.

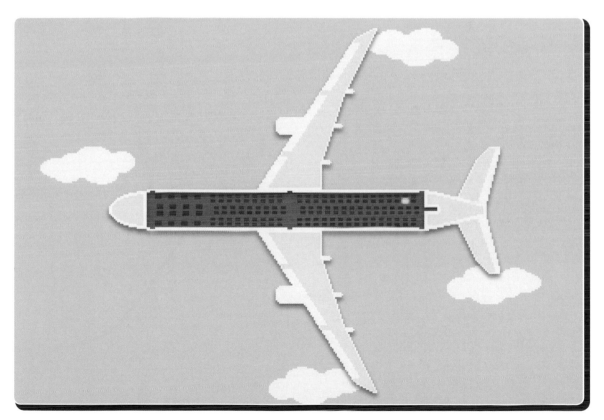

SHOULD YOU CHANGE LANES IN A TRAFFIC JAM?

You've got places to go and people to see. You're already running a little late, but that's fine because you're sailing along - even the traffic lights stay green as you approach. But then the inevitable happens: you hit a jam. Bumper to bumper, the tail-lights stack up as you grind to a halt.

Looking around, you realize that the lane next to you seems to be flowing more smoothly. Tempted, you switch lanes. According to a study published in *Nature* in 1999, many of us would do the same. In a sample of 110 drivers, forty-six per cent said they would consider moving over if the next lane appeared to be moving ten per cent faster. If it was moving fifty per cent faster, then seventy-two per cent would switch. But does switching lanes actually get you through the traffic any faster? Or are you about to join another queue of vehicles whilst the one you were in frees up?

The answer seems to be that it depends. Back in 2010, Alexander Lobkovsky Meitiv tackled this question on the *Playing With Models* blog. His work is based on a mathematical model of traffic formulated by Kai Nagel and

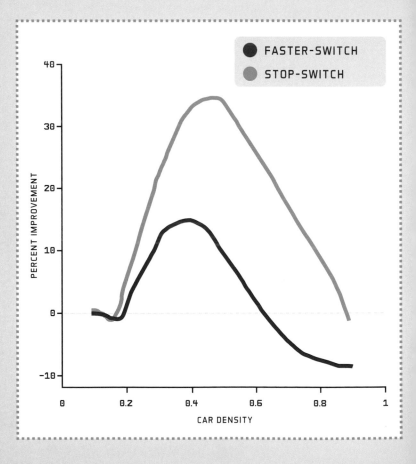

Michael Schreckenberg, published in the *Journal de Physique* in 1992.

Meitiv looked at two lane-switching scenarios. The first – which he calls "stop switch" – involves changing lanes if the car in front of you has stopped but there is space in an adjacent lane. The second – "fast switch" – means changing lanes even if the car in front is moving, so long as the traffic is moving faster in the lane next door. He found that the "stop switch" tactic offers a greater improvement in average speed than "fast switch". However, he also concluded that the extent to which either tactic is beneficial depends on the overall density of the traffic (*see graph*). In fact, employing either tactic in very heavy traffic will cause you to slow down rather than speed up. So be careful. But, used wisely, the "stop switch" tactic can boost your average speed through moderately heavy traffic by around thirty-five per cent.

However, let's keep this little nugget of information to ourselves. That's because it turns out that if everyone employs this tactic then it causes everyone to slow down. It will only be advantageous to you if only a few people are doing it. So mum's the word.

This switching method was backed up by two experiments on the *Mythbusters* television show. In the first, two cars drove from San Fransisco to San Jose. One always stayed in lane, but the other driver was free to choose whichever they deemed to be the optimal lane. The lane-switching car arrived first. In the second experiment, five cars were involved in driving down a four-lane highway. Cars one to four had to stick in their respective lanes, but the fifth car was free to roam. The driver that could change lanes arrived before all the others – four per cent faster than the second car to arrive and twenty-five per cent faster than the last.

However, a word of caution: whilst switching lanes may get you to your destination faster under certain circumstances, repeated lane changing also brings increased risk of an accident. So, always be careful and weigh up the time gained against the elevated danger level.

HOW TO BUILD THE BEST SANDCASTLE

No trip to the beach is complete without dabbling in a little sandcastle construction – even if the incoming tide eventually washes away your lovingly crafted creation. But what exactly goes into building the ultimate sandcastle?

If you've ever built a sandcastle, then you'll know that one of the key factors to success is getting the right sand-to-water ratio. Too little water and the grains won't stick together sufficiently, but too much and you are left with just a sandy sludge. Two studies, published in *Nature Physics* in 2005 and in *Nature Mechanics* in 2008 respectively, suggest the droplets of water act as tiny liquid handshakes – or "capillary bridges" – holding the sand grains together. So, getting the optimal ratio is crucial. That second study found that eight buckets of sand for every one bucket of water should generally do the trick.

However, physicists from France, the Netherlands and Iran later teamed up to look into the properties of sandcastles, publishing their results in the journal *Scientific Reports* in 2012. They found that to build a great sandcastle you should use a lot less water – just one bucket of water for every fifty buckets of sand. This allowed the researchers to build a sand tower measuring only sixteen centimetres (six inches) across but towering to a height of 1.2 metres (four feet).

The other key to the scientists' success was compaction. Packing the sand tightly together increased its strength by up to thirty per cent. So, while it is fine to use a bucket and spade to make sure the quantities are right, you shouldn't use them to build the actual sandcastle. Packing the sand together with your hands will lead to a more compact building material. You might think that you can compress it tightly into a bucket, but it is when you try and remove the bucket that a lot of your hard work is undone. The common practice of hitting the bottom of the bucket with a spade should also be avoided, as this can fracture the compact structure of sand that you're trying to create. The professionals use mechanical thumpers to get it right.

By filling cylindrical PVC tubes of different diameters with sand before removing them to reveal sand towers, the research team could also look at the relationship between a sandcastle's height and width. Unsurprisingly, they found that the wider the base of the sandcastle, the higher they could build it without it collapsing. However, their real insight was the exact relationship between the two. Tripling the width of your base will allow you to approximately double the height of your sandcastle (see graph). The world record for the highest sandcastle is currently 13.97 metres (46 feet), set in October 2015. However, according to the science, the constructors could actually have built an even taller structure, given the size of the base that they used.

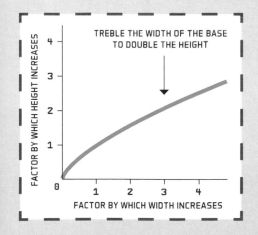

FACTOR BY WHICH HEIGHT INCREASES

TREBLE THE WIDTH OF THE BASE TO DOUBLE THE HEIGHT

FACTOR BY WHICH WIDTH INCREASES

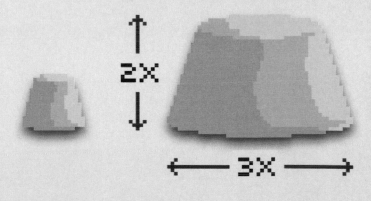

2X

3X

CHAPTER 7

MONEY

HOW TO SAVE MONEY IN THE SUPERMARKET

There was a time when we humans were hunter-gatherers. In order to survive, we had to track and kill other animals and forage around for fruit to sustain us. Now, we just have to drive a few miles to a big building with aisles crammed full of everything we need.

However, handing over the job of supplying our food to someone else means that we have become more susceptible to being "taken for a ride". After all, supermarkets are big business – the more we spend inside, the more profit they make.

So, we want to arm you with all the tips and tricks you will need to make sure that you leave the supermarket with as little damage to your bank balance as possible.

The first sure-fire way to rack up a substantial shopping bill is to "impulse buy". As we will see, supermarkets employ many tactics to tempt us into buying as much as possible. So, the first thing to do is go into the shop with a carefully prepared list. Check your cupboards and fridge to see what you already have and plan out your meals for the week. Then you can write a shopping list of all the things you want. Sticking to the list means you are less likely to be tempted by stuff you don't need when you are in the store.

In addition, a pre-prepared list means that you will only have to visit the aisles containing the items on that list. This is important because, according to a study by the Marketing Science Institute, shoppers who visited most or all of the aisles in the shop ended up checking out with sixty-eight per cent more unwanted items compared to those visiting fewer aisles. It is why everyday items such as bread and milk are often at the back of the store, forcing you to pass a lot of other temptations to get to them. The same study found that it is best to go shopping alone – the amount of unwanted items goes up by eight per cent if you group shop. Even your choice of trolley has a distinct influence on how much you purchase. Bigger trolleys tempt you into filling them up with more, but going for a small basket isn't necessarily the solution either. A 2011 study published in the *Journal of Marketing Research* found that the strain of carrying a heavy basket leads us into more impulse buys – such as candy and soda – as "reward" for the effort. Small trolley it is, then.

You should take your own music to the supermarket, too, as in-store tunes have been shown to affect our purchasing habits. In 1999, researchers from the University of Leicester, in the UK, published the results of their study in the *Journal of Applied*

THE AMOUNT OF UNWANTED ITEMS GOES UP BY 8% IF YOU GROUP SHOP.

DOS.

- ✓ Shop online.
- ✓ Take headphones.
- ✓ Use a small trolley.
- ✓ Make a list.
- ✓ Look at items on the top and bottom shelves.

DON'TS.

- ✗ Buy only from the middle shelf.
- ✗ Fall for the illusion of red stickers.
- ✗ Use a big trolley or a basket.

Psychology. They found that German wine outsold French wine when German music was playing, and vice versa. A subsequent survey of shoppers found that customers were unaware that music can influence their choices. Ambient music can also have an effect on the speed at which we move around the aisles. According to a 2000 study published in the *Journal of Business Research*, shoppers exposed to unfamiliar music spent longer in the store than those who heard familiar music. So, by putting your headphones in you aren't being influenced on what to buy, and by listening to your own music you'll spend less time exposed to temptation. Plus, you get around the tactic employed by some supermarkets of playing slow music in areas where they want you to linger (for example, where the premium items are to be found).

So far we have looked at ways to avoid buying things you don't really want. But how do you go about not overspending on items that are actually on your list? Well, first off, beware the red label – especially if you are a man. A 2013 study, by researchers at Drexel University in Philadelphia, found that both men and women believed they were saving more money when the price was written in red as opposed to black. However, for men the effect was much more pronounced (*see graph*).

It may sound simple, but it also pays to look on all the shelves in a particular aisle. There is a saying within the retail industry: "eye level is buy level". You'll often find that the most expensive items are between eye and waist level – a height that means you can get at them more easily. So, try bending down to the bottom shelf or reaching up to the top one and you

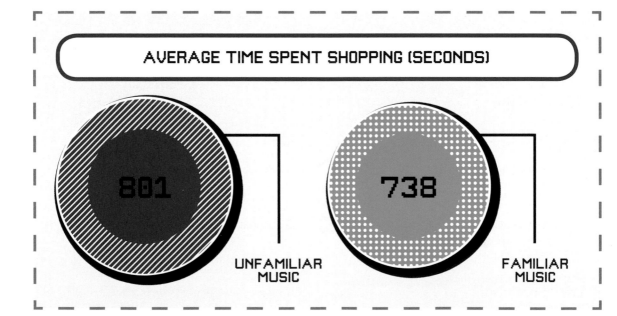

AVERAGE TIME SPENT SHOPPING (SECONDS)

801

738

UNFAMILIAR MUSIC

FAMILIAR MUSIC

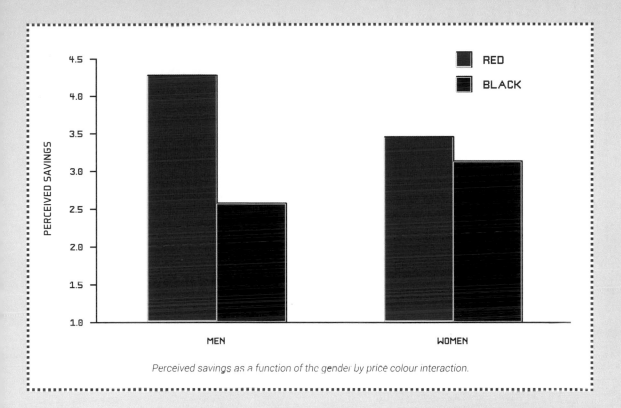

Perceived savings as a function of the gender by price colour interaction.

will probably find that the items there are cheaper. Unless, of course, you are in an aisle containing products aimed at children. Here, lucrative items are placed lower down so as to attract the attention of children, who the retailers hope will then pester their parents into buying them. A team of researchers at Cornell University examined the angle at which cartoon characters on cereal boxes were looking. They found that on average they were looking down at an angle of 9.67 degrees. For people depicted on adult-orientated cereal boxes, this angle was close to zero. The reason? Brands want us to make eye contact with the people and characters on packaging, and children are smaller, so therefore have to look up to the products.

Exactly where we look on shelves was investigated further by the consumer watchdog Which? in a 2014 experiment. They attached eye-tracking equipment to volunteers who were given lists and asked to go shopping. The guinea pig shoppers all ended up with impulse buys. The study found that we read supermarket shelves like a book – from left to right. So, if retailers increase the price gradually from left to right, we don't notice the overall big difference between the cheapest and most expensive item on a shelf.

Of course, a sure-fire way to avoid all of these problems is to shop online from the comfort of your sofa and never enter the supermarket in the first place. According to a 2014 poll of online shoppers by eDigitalResearch, only seven per cent of the 1,154 people surveyed said they made more impulse purchases online than in store. However, if you do have to venture to the shops, being aware of these simple tricks could mean all the difference between overspending or returning with a little more cash in your pocket.

HOW TO BE A
BETTER HAGGLER

It seems that the art of haggling is in decline, at least according to a 2016 British study on the purchase of new cars. The survey – carried out by motoring magazine *AutoTrader* – found that fifty-six per cent of buyers paid the full asking price, without haggling at all.

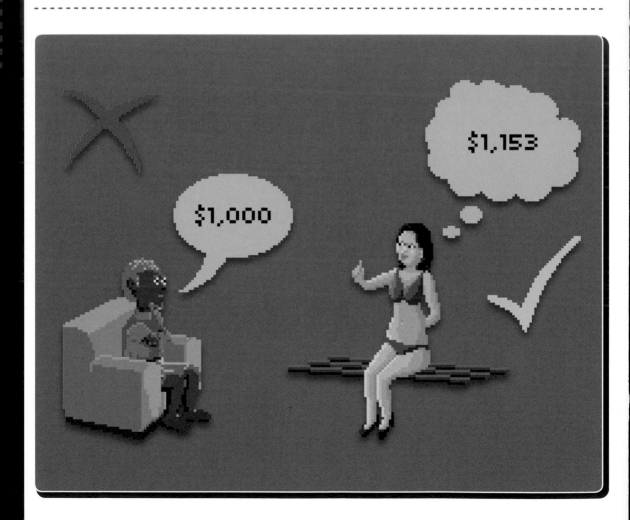

That is a twelve per cent increase compared to the previous year. The difference between the generations was particularly stark — of those aged seventeen to twenty-four, sixty-seven per cent paid the full price, compared to just thirty-five per cent of those aged fifty-five to sixty-four.

There are many situations in which negotiating on price can get you a better deal, but many of us are just not confident in our haggling abilities. Never fear, here at Geek Towers we've compiled a list of top techniques to turn you into a better haggler.

One of the key decisions to get right is your opening offer but, according to 2013 research published in the *Journal of Experimental Social Psychology*, most people get it wrong. The study, led by Professor Malia Mason from Columbia Business School in New York, looked at negotiations between 1,254 students over fictitious items such as jewellery and used cars. The results of the negotiations were read out in class, so as to give the participants a real-life incentive for success and ensure a realistic exchange. The most successful hagglers employed an unusual tactic — starting with an opening offer that wasn't a round number (*see graph*).

Let's say the sticker price of an object is $1,000. Most of us trying to haggle would counter-offer with a lower, round number — say $800. But Mason's study found that picking a more jaunty number such as $813 led to greater success. Mason believes this is because a more pinpoint figure gives the impression of knowledge — it appears as if you must have done your homework in order to arrive at such an accurate value. Even if you haven't, your "opponent" is less likely to mess with you.

It also seems that our surroundings may play a part in how entrenched we become in our negotiating positions. In 2010, psychologists from Harvard, MIT and Yale universities published some intriguing findings in the prestigious journal *Science*. Their experiment saw eighty-six participants involved in mock negotiations over the price of a new car. Some volunteers were seated in hard seats, the others in

soft ones. Those seated in the hard chairs were more rigid in their negotiation, with smaller gaps between successive counter-offers. It appears the driving seat in haggling is a firm one.

Finally, it seems that an attractive woman can have a negative effect on the haggling abilities of men with high levels of testosterone. Back in 2006, Belgian researchers Bram Van den Bergh and Siegfried Dewitt published their findings in the journal *Proceedings of the Royal Society B*. The male participants in their study were asked to play a game in which they had to agree how to split a fixed amount of money between them. The men with the highest testosterone levels were the most fiendish negotiators, but their powers began to wane when they were shown pictures of models in bikinis or even when they handled a bra. However, images of elderly women or landscapes, or holding a T-shirt, had no such effect. Make of that what you will!

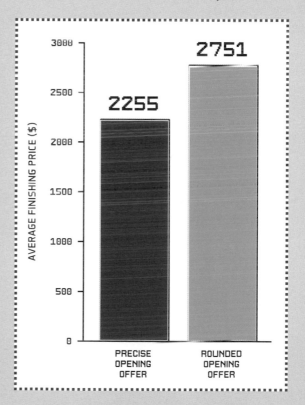

HOW TO BE A BETTER SAVER

What do you do with your money? Are you a spender or a saver? It seems that as we age, many of us wish we had saved more. According to a 2013 survey, fifty-one per cent of those aged forty or above regretted not squirrelling more money away when they were younger.

However, consistently saving money seems to be a perpetual struggle for some. If this is you then it may not be entirely your own fault. According to one study, our genes play a major factor in how disciplined we are with our cash. The research, published in the *Journal of Political Economy* in 2015, looked at twins in Sweden and concluded that genetic factors account for about a third of our financial behaviour.

Fortunately, even if you lost out on the genetic lottery, there are still things you can do to improve your chances of accumulating a decent nest egg. The first thing to consider is having a single goal rather than multiple savings targets. A 2011 study

in the *Journal of Marketing Research* looked at people from many different backgrounds, including some from rural households in India, middle-income Canadians and professionals in Hong Kong. The researchers from the University of Toronto concluded that multiple savings goals typically led to trade-offs between them and therefore an increased likelihood that people will defer saving while they try and work out what to do (*see graph*).

That said, thinking of time in a linear fashion like this might not be the way to go at all. The conventional way many of us tackle saving is to picture some future event, set ourselves a goal and

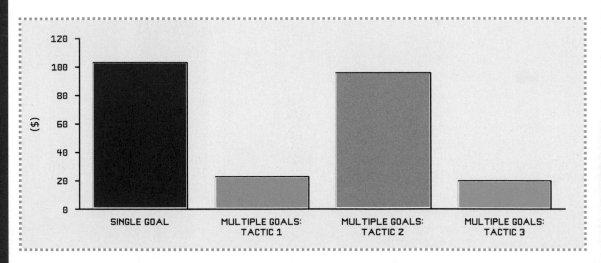

work towards achieving that. We focus more on the end target than the act of saving itself. And yet, a 2013 study led by Leona Tam and published in the journal *Psychological Science* found that thinking of time in a more cyclical way led to greater savings success. Participants were split into two groups – "linear savers" and "cyclical savers". The first group were given instructions on a particular method of saving, including:

"This approach acknowledges that one's life is made of separate and progressive time compartments such as the past, present and future. We want you to think of the personal savings task as part of such a linear progress."

Whereas the cyclical group were told:

"This approach acknowledges that one's life consists of many small and large cycles, that is, events that repeat themselves. We want you to think of the personal savings task as one part of such a cyclical life ... just focus on saving the amount that you want to save now, not next month, not next year."

For the next two weeks the participants were asked to save money according to these approaches and the results showed a marked difference. Those in the linear group saved an average of $140, compared to $223 for those following the cyclical mindset.

Finally, your willingness to save can be influenced by how you think about the timescale involved. A 2015 study, also published in *Psychological Science*, found that participants were motivated to start saving four times sooner when they thought about some future event in terms of days and not years. Clearly, seeing your retirement as 10,950 days away rather than thirty years is enough to inject a sense of urgency...

HOW TO SPLIT THE RESTAURANT BILL

You know the drill. You're out dining with a group of friends to celebrate an important milestone. The food is delicious, the wine is flowing and the conversation couldn't be better. Everyone is having a really great time. But then the inevitable happens - the atmosphere of the whole evening is shattered when the bill comes.

Someone pipes up that you should all split it evenly between you. But then another person counters with the fact that they've only had a salad and no dessert. Others haven't been drinking and are reluctant to subsidize the intoxication of others. It's a familiar situation, immortalized in the second season of the hit US sitcom *Friends* ("The One With Five Steaks and an Eggplant").

According to a study published in the *Economic Journal*, splitting the bill evenly drives up the price for everyone. The researchers took groups of six people out for dinner in a restaurant. The diners either split the bill according to what they ate, split it evenly between them, or the entire meal was paid for by the research team. People ordered more when splitting evenly compared to when paying individually (an average of 1.87 items to 1.67). The average cost

to each person when splitting equally was $50.90, compared to $37.30 for paying only for what they consumed. The logic behind these decisions was probably something along the lines of "I'll order the lobster because I won't be paying for all of it." But you're not the only one at the table, and it seems many of your fellow diners might have had the same brainwave. Alternatively, it could reflect peoples' determination to get their money's worth. "If Joe is ordering the fillet steak, then I had better order something expensive too, so I'm not getting a raw deal."

Interested in the motivation of diners under the even-split scenario, the researchers devised a fourth experiment. This time the six participants were told that they would only pay for one-sixth of what they ordered, with the rest of the tab being picked up by the researchers (rather than their fellow diners). Given this set-up, people ordered an average of 2.08 items at a total average cost of $57.40 — more on both fronts than the even-split scenario. Maybe people limit how much they take advantage of the other people at the table, even though they still do so to some degree.

Given these results, it is odd that we ever choose to split the bill evenly. The study backs this up, finding that the majority of us don't even want to. Diners were asked before the experiment how they would prefer to split the bill, and eighty per cent opted to pay only for what they personally ordered. Yet they still ordered more when they found out that the bill was to be split evenly.

However, dividing the bill up to the penny is time consuming. So, is there a reasonably fair, but quick way to divide up the check without poring over every last cent? Stand up mathematician Matt Parker, who believes there is, devised what he calls "Standard Meal Units". One main course and two drinks represents one unit. Those having salad only would clock up 0.6 units, with heavier drinkers amassing 1.4 units. Add up everyone's total units and divide the total bill by this number, before multiplying by everyone's individual unit total.

ALICE = SALAD = 0.6

BOB = MAIN MEAL + TWO DRINKS = 1.0

CHARLIE = MAIN MEAL + MORE THAN
TWO DRINKS = 1.4

TOTAL BILL = $63

PRICE PER UNIT = $50/3 = $21

ALICE'S COST = 0.6 X $21 = $12.60

BOB = 1 X $21 = $21

CHARLIE = 1.4 X $21 = $29.40

HOW TO SPEND LESS IN RESTAURANTS

A menu is simply a list of what a restaurant offers, right? Wrong - it's their chance to sell to you. A good menu is a carefully crafted, handheld advertisement designed to get as much money out of you as possible.

Take something as small as the way the prices are written. If you eat out regularly you might have noticed that restaurants are increasingly dropping the $, £ or € symbols from menus. That's no accident. A 2009 study by Sybil Yang at Cornell University found that diners racked up higher bills when the menu displayed prices without a dollar sign in front of them.

How a dish is described has a great effect on consumers' choices, too. In a separate Cornell study, published in the *International Journal of Hospitality Management*, researchers played around with the wording of certain dishes. "Seafood Fillet" was changed to "Succulent Italian Seafood Fillet" and "Red Beans and Rice" was swapped for "Cajun Red Beans and Rice". Sales of the two dishes rocketed by twenty-eight per cent, despite no actual changes to the food taking place. The researchers also found that diners were willing to pay twelve per cent more for the dish when it had the fancier name.

Then there is "price anchoring". This is the practice of putting a really expensive item right next to the thing they really want you to order (often the one with the biggest mark-up). Throw in a $70 lobster dish and suddenly the other items around it look like better value for money. Placing these anchor items in boxes or a bold font ensures our eyes are drawn to them first. A similar principle works in reverse for the wine list. In order not to look tight-fisted, diners often go for

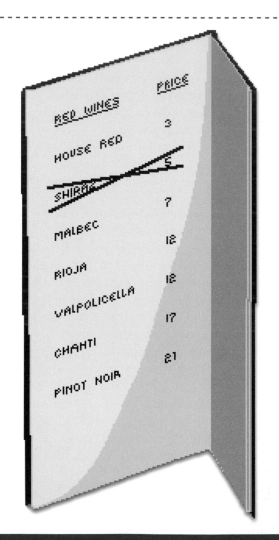

the second cheapest option. However, for precisely that reason, restaurants often place the bottle with the biggest mark-up there. You would probably be better off going for the cheapest bottle on the list.

Restaurateurs can also dictate how long we dwell in their establishment – and therefore how much we spend – by the type of music that they play. A study by British researchers published in the journal *Environment & Behaviour* found that customers spent an average of £32.52 on food and drink when classical music was playing, compared to £29.46 when pop music was blaring out. Smell can have an effect, too. Another study published in the *International Journal of Hospitality Management* found that customers spent more when a lavender smell was present than when they could smell a lemon scent or no scent at all.

Watch out for the staff trying to extract as big a tip out of you as possible, too. You think that mint on the plate with the bill is just a nice gesture? In truth, it is a bribe. A study published in the *Journal of Applied Social Psychology* found that small gifts such as candy with the bill meant that diners gave higher tips. Research also suggests that we tip more if the server handwrites a "Thank You" or draws a smiley face on the bill. We're also more generous if the waiter kneels down beside us when taking our order. Oh, and beware blonde servers. A study published in the journal *Archives of Sexual Behaviour* in 2009 says we tip them more.

Armed with this knowledge, you should be able to save money just by being aware of some of your unconscious biases.

FOOD SPEND: CLASSICAL VS POP MUSIC

VARIABLE	CLASSICAL MUSIC	POP MUSIC
STARTERS	£4.92	£4.84
MAIN COURSE	£14.72	£14.52
DESSERT	£3.42	£2.55
COFFEE	£2.07	£0.00
BAR	£3.51	£3.06
WINE	£4.88	£4.49
TOTAL DRINK	£8.39	£7.55
TOTAL FOOD	£24.13	£21.91
TOTAL SPEND	£32.52	£29.46

HOW MUCH MONEY DO YOU NEED TO BE HAPPY?

The Beatles famously sang about the fact that money can't buy you love, but what about happiness? Is how content you feel linked to the size of your bank balance?

The answer it seems is yes – and no. A famous study carried out by Nobel prizewinners Angus Deaton and Daniel Kahneman in 2010 analysed data from surveys covering 450,000 Americans. Questions included rating how happy they were the previous day and if they thought they were living the best possible life for them. Respondents also disclosed how much they got paid. Deaton and Kahneman's conclusion, published in the journal *Psychological and Cognitive Sciences*, was that the magic figure is an annual salary of $75,000. Up to this point overall contentment

with your lot in life increases in line with earnings, yet any more money than that made no further difference.

However, the duo was quick to point out two different forms of happiness – a very nebulous concept that is notoriously hard to measure. On the one hand, there is your day-to-day happiness; on the other your general contentment with your life as a whole. Deaton and Kahneman found that the $75,000 threshold applies to the latter but not the former. So, having more money isn't going to make you feel more cheerful in the morning...

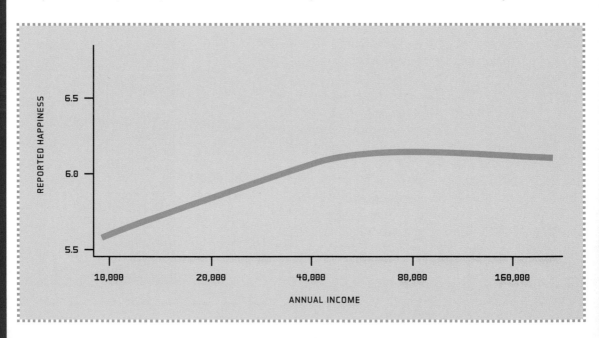

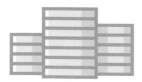

One way to test how a lot of money affects your happiness is to study lottery winners – people who have been catapulted into a sudden new level of wealth. A famous paper from 1978, published in the *Journal of Personality and Social Psychology*, found that a big jackpot win did cause a sudden spike in happiness, but that it wore off pretty quickly. This forms part of an idea called the "hedonic treadmill" – that, as our wealth increases, so do our expectations, and so there is no net gain in our happiness. We return to a sort of happiness "set point". Sure, $1 million would be nice, but very soon you adjust to that amount of money and start coveting what you could do with $5 million.

According to some studies, this set point might have less to do with a financial windfall and more about hitting the jackpot in the genetic lottery. In the mid-1990s, behavioural geneticist David Lykken studied fifteen hundred pairs of twins and published his results in the journal *Psychological Science*. He found that life circumstances – including salary – accounted for only two per cent

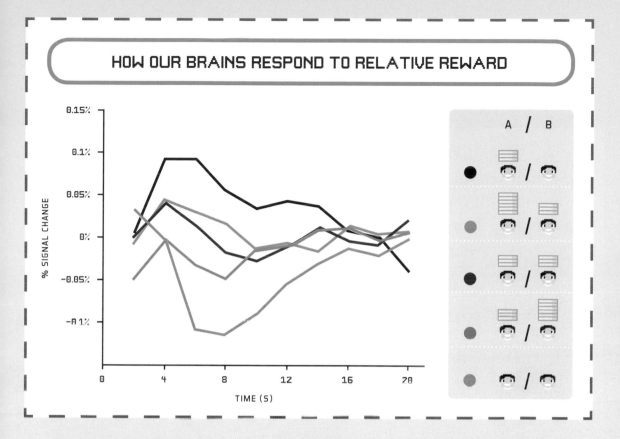

HOW OUR BRAINS RESPOND TO RELATIVE REWARD

of the difference in happiness between identical twins (those with matching genes). Twins with high incomes were no happier than their less well-off sibling. He also concluded that the level of our happiness set point is half-determined by the genes that we inherit from our parents and half by our day-to-day experiences.

If our absolute wealth does not make us happy, then could our relative wealth compared to something – or someone – else have an effect? In 2007, neurologist Christian Elger published a paper in *Science* in which he reported findings from MRI brain scans on adult men (*see graph*). Pairs of participants played a game side by side in which they had to count dots on a screen and were offered varying financial rewards for success. Meanwhile, the researchers studied which areas of their brains were

activated. They found that activation in the brain's "reward centre" was stronger when one player got more money than the other compared to when they were both paid the same amount. This importance of relative wealth was backed up in 2010 by a study in *Psychological Science* led by Christopher Boyce. His work showed that happiness is linked to how your wealth is ranked compared to your peers, not how much money we make in absolute terms.

So, it seems that once your basic needs are met, the total amount of money you earn has only a marginal effect on your overall happiness. But that doesn't mean money can't buy you happiness if you spend what you do have in the right way. Fortunately, we've found other methods in which to maximize the joy that money can bring, so that you get the biggest bang for your hard-earned buck...

SPEND MONEY ON EXPERIENCES, NOT THINGS

We live in a very materialistic culture, in which the latest must-have items act as a form of social currency. Cutting-edge gadgets, cars and fashions can cause us to go into a frenzy, with some queuing up for hours to get their hands on them. But can buying stuff really make you happy? According to a growing body of psychological evidence, spending more of your hard-earned cash on experiences rather than material things is a surer path to contentment.

As far back as 2003, a paper entitled "To Do or to Have? That Is the Question" was published in the *Journal of Personality and Social Psychology*. The study, by Leaf Van Boven and Thomas Gilovich, surveyed a cross-section of the American population, asking participants to rate how happy recent purchases of over $100 had made them. Fifty-seven per cent of respondents rated experiences as providing a greater level of happiness than material purchases. That compared to only thirty-four per cent ranking material purchases over experiential ones. The difference was particularly marked among students and homemakers (sixty-seven per cent to twenty-five per cent).

DEMOGRAPHIC CATEGORY	TYPE OF PURCHASE	
	EXPERIENTIAL	MATERIAL
AGE		
21–34 (350)	59%	36%
35–54 (645)	58%	31%
55–69 (268)	49%	38%
EMPLOYMENT		
EMPLOYED FULL- OR PART-TIME (941)	58%	33%
RETIRED OR UNEMPLOYED (218)	47%	39%
STUDENTS AND HOMEMAKERS (102)	67%	25%
GENDER		
MALE (591)	51%	38%
FEMALE (672)	62%	30%
RESIDENTIAL ENVIRONMENT		
URBAN (363)	56%	35%
SUBURBAN (654)	59%	31%
RURAL (246)	49%	40%

Then, in 2014, Gilovich teamed up with Amit Kumar and Matthew Killingsworth to publish "Waiting for Merlot: Anticipatory Consumption of Experiential and Material Purchases" – a study in the journal *Psychological Science* – examining the spending habits of more than two thousand adults. The researchers got in contact with the participants at random times using a smartphone app, and found that they were thinking about a future purchase on nineteen per cent of occasions. Participants were also asked to rate their happiness and excitement at the time they were contacted on a scale of 0 to 100. Crucially, the study found that participants waiting for an experience derived significantly more happiness and excitement than those anticipating a material purchase. Buying stuff can still make you happy – it is just that buying experiences seem to make you happier.

So, it is both the build-up to and reflection on experiential purchases that appears to contribute more to overall happiness. This theory was explored further by Kumar and Gilovich – by now noted for their paper titles – in "We'll Always Have Paris", published in *Advances in Experimental Social Psychology*. The pair reviewed much of the research in this area and found that experiential purchases tend to be more socially rewarding, more closely tied to people's identities, and thought about and enjoyed more on their own terms and less comparatively.

In particular, it seems it is talking about our experiences after the event that keeps the happiness boost rolling. The novelty of a new television or smartphone wears off pretty quickly, but by telling others about the awesomeness of our latest trip, or how off the charts that concert was, we are reliving that experience all over again. So, next time you have to sit through yet another set of someone else's holiday snaps, remember that it is probably making them happier.

WHY YOU SHOULD GIVE MONEY TO CHARITY

So, we know that buying experiences over things is likely to make you happier in the long term, but what about spending money on others? Sure, giving to charity is a worthwhile endeavour - even if it has no net benefit to you - but what if you knew that donating some of your cash to a worthy cause gives you the same kick of pleasure as sex or eating chocolate? You'd be more likely to give, right?

Well, that's exactly what scientists found when they looked into what's going on in our brains when we give money away. Researchers at the University of Oregon placed nineteen women in an fMRI machine for an hour. Initially, the participants were given $100 and told they could keep whatever was left once the time was up. However, during the experiment their balance, which was displayed on a computer screen, changed. Sometimes this was because a portion of their cash was donated to a local food bank, but at other times money was added to their pot. Crucially, some of the food bank donations were at the discretion of the participant – they could say "yes" or "no".

Led by William Harburgh, the team published their results in the prestigious journal *Science* in 2007. They reported that seeing money go to a good cause lit up the same pleasure centres in the brains of participants as when they received extra money. Those are the same reward centres that are activated when we eat chocolate or have sex. The activation was even greater when the participants voluntarily donated their money (although they did reject more than half of the voluntary payments).

This idea was further backed up a year later when Elizabeth Dunn from the University of British Columbia published more research in *Science*. She asked 632 Americans to disclose their average monthly expenditure, along with rating their own happiness. People who spent more money on others came out happier. Of course, that doesn't necessarily mean that giving money away induces happiness. Perhaps people happier for another reason are more likely to be generous?

To further tease out a link, Dunn performed a second study, in which she looked into how employees spent a company bonus. Two months after they had received the windfall, Dunn asked them how they had spent it. Those who splashed cash on more altruistic endeavours, rather than just themselves, faired better on the happiness front. Dunn's final experiment was to give forty-six participants either $5 or $20, asking half of them to spend it on themselves and the other half to either buy something for someone else or give it to charity. Those that gave the money away exhibited a greater leap in happiness that evening, compared to the morning.

This effect isn't just limited to adults, which suggests it might be a universal quality of human interactions. In a 2012 study published in the journal *PLoS ONE*, Dunn and colleagues observed children as they were asked to give away to an interactive puppet one of the eight treats they had been given. Meanwhile, volunteers were rating the toddlers' facial expressions for happiness. They found that the children were happiest when they gave away one of their treats, even more so than when they first met the puppet or were first given their pile of treats.

So, if happiness is what you're after, then it seems you could do a lot worse than give some of your disposable income away to your favourite cause.

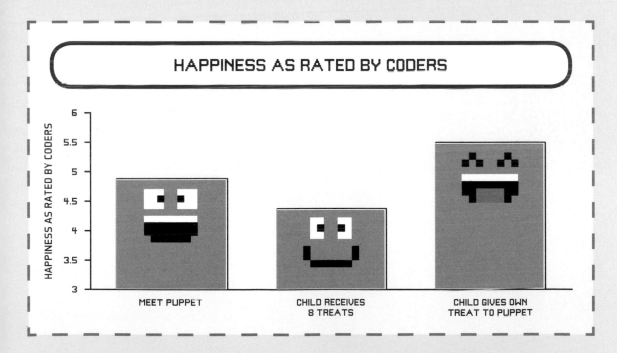

HAPPINESS AS RATED BY CODERS

CHAPTER 8

TECHNOLOGY

THE BEST STRATEGIES FOR BUYING AND SELLING ON EBAY

Online auctions are still incredibly popular. The market leader eBay has more than 160 million registered users, who conduct more than 250 million daily searches through 800 million newly listed items.

With such stiff competition, buying and selling successfully on the platform can seem to be a bit of an art. In fact, it is just as much a science. We've been digging through the research to bring you the top tips to ensure eBay success.

Unless you are bidding for a very niche item, there are likely to be several versions of the product you are after listed on the site. Picking which one to battle over could mean the difference between paying over the odds and walking away with a bargain. If you're happy that two items are essentially the same, one thing to look at is the background colour used on the listing. Results published in the *Journal of Consumer Research* in 2013 examined eBay auctions for Nintendo Wii bundles and found that users bidding on a listing with a red background made higher bid jumps and generally bid more aggressively. So, if you're buying, avoid bidding on items with red backgrounds. Conversely, if you're selling, consider photographing your item in front of something red.

Another longstanding conundrum is when to bid. Should you get in early but risk driving up the price? Or swoop in at the last minute to steal the win from under the noses of your rivals? South Korean mathematicians Byungnam Kahng and Inchang Yan have extensively investigated this issue, studying over half a million auctions across both the US and Korean eBay sites. Their analysis, published in the journal *Physical Review E*, found that bidding activity ramps up towards the end of an auction and their advice is simple: wait until the very last minute, a method known as "sniping". If you don't trust yourself to get the bid

TOP TIPS:

BUYING.

 Don't bid on items with red backgrounds.

✓ Use sniping software to bid at the last minute.

✓ Negotiate on offers with round starting prices.

SELLING.

✓ Use a red background on your auction listings.

✓ For maximum sale price, use non-round starting prices.

✓ For quick sales, use round starting prices.

in on time, you could use one of the many free online sniping tools that will automatically bid for you up to your maximum amount. This finding backs up 2002 research published in the journal *American Economic Review* by economists Al Roth and Axel Ockenfels.

Yet while eBay is most associated with auctions, the site is increasingly leaning towards negotiation as its main selling method. A seller will list an item at a certain price, only for potential buyers to try and knock them down through a series of offers. According to researchers Matthew Backus, Thomas Blake and Steven Tadelis, there is a marked difference in the outcome if the initial asking price is a round

figure. Specifically they found that Buy It Now prices that were a multiple of $50 consistently attracted opening counter-offers that were between five and eight per cent lower than those listings with more random starting values. So, the take-home message is that if you list an item for $98 rather than $100, you're more likely to have extra money in your pocket at the end of the negotiation process. However, if you care more about a quick sale than the highest possible price, you might consider sticking with the round numbers. Those listings were found to end six to eleven days sooner and were three to five per cent more likely to sell.

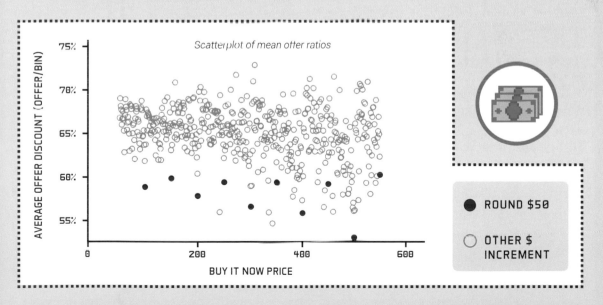

Scatterplot of mean offer ratios

AVERAGE OFFER DISCOUNT (OFFER/BIN) — BUY IT NOW PRICE

● ROUND $50
○ OTHER $ INCREMENT

HOW TO STOP YOUR EARPHONES TANGLING

It surely has to rank as one the most frustrating aspects of modern technological life. Whilst your earphones allow you to listen to your favourite music on the go, they seem to have a life of their own...

Pull them out of your pocket and they've suddenly managed to contort themselves into a series of impenetrable knots. You then spend the next ten minutes fiddling around and trying to free them from their entanglement. If you're doing this every day, then you're spending the equivalent of a day and a half every year on this infuriating task.

This knotting happens incredibly quickly, according to a study published in the journal *Proceedings of the National Academy of Sciences of the United States of America (PNAS)*. The authors of the research – Dorian Rayner and Douglas Smith – found that complex knots can form within seconds. They put strings in a box and jumbled the box around for ten seconds. This experiment was repeated 3,415 times and the results showed the strings knotting in one hundred and twenty different ways and crossing themselves up to eleven times. Unsurprisingly, they found that longer, less stiff cables – just like earphone wires – were more likely to knot (*see graph*). Don't get us *started* on Christmas tree lights!

So how do you prevent it from happening? The Internet is awash with suggested methods to prevent this scourge, but many of them involve intricate and complicated ways of tying the wires so that they don't knot. Here at Geek Towers we prefer the method discovered by Aston University physicist, Robert Matthews. He enlisted the help

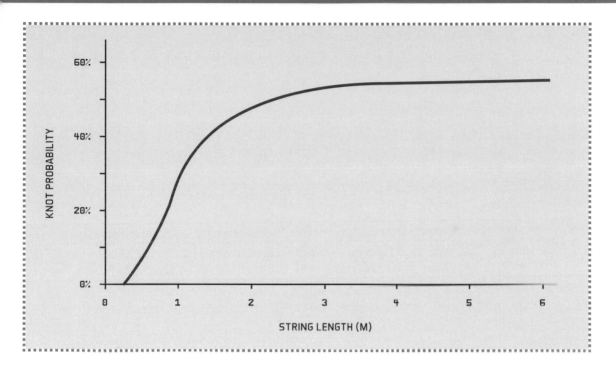

KNOT PROBABILITY

STRING LENGTH (M)

of pupils at a local school, who performed twelve thousand experiments with parcel string varying in length from fifty-five to 183 centimetres. Their results, published in the journal *School Science Review,* illustrate how joining the ends of the string together reduced the risk of knotting by a factor of ten. Applying this "loop conjecture" to earphones, Matthews suggests simply clipping the wires together below the ear buds. You can reduce the risk of knotting even further by then clipping the audio jack to the other wires just above the previous clip. You could use a small bulldog clip or a large paper clip, or buy a clip designed for the job. Either

way, a quick clip and unclip has to be better than fiddling around with fiendish knots.

This knot research might seem a little off the wall, but it turns out it could have more profound consequences than keeping your earphones straight. Our DNA, for example, takes the form of long threads wrapped around each other in a famous double helix shape. For DNA to replicate, it first has to untangle itself. So, work on knot formation might provide insights into what happens when this process goes awry. Similarly, understanding why cords can knot could help prevent umbilical cords from tangling around unborn babies.

WHAT'S BETTER: BOOK OR E-READER?

Technology for technology's sake doesn't always cut it. Sometimes consumers rebel against the relentless march of the latest gadgets in favour of a more traditional approach. For example, take the recent surge in sales of vinyl records. Retailers are reporting sales figures for these "LPs" (long players) that haven't been seen since before the CD (compact disc) really let fly in 1990. Given that you can stream the same songs in an instant online, it says something about the desire of many to have a physical object to hold in their hands.

The same can be said of the battle between e-readers such as Amazon's Kindle and traditional books. Growth in the former has quietened down in recent years, with sales of physical books back on the increase. But which is the better option?

E-readers certainly have their advantages. They are often lighter and more compact, and can store multiple books at once. Plus, according to environmental consulting firm CleanTech, as long as you read over twenty-three books on your e-reader, it is better for the environment, as well.

But does how we consume our books change our experience of reading them? Researchers Anne Mangen and Jean-Luc Velay believe it does. They

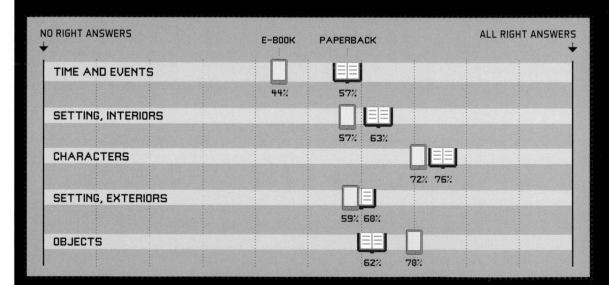

took fifty graduate students and split them into two groups. One group was asked to read a passage from an Elizabeth George novel on a Kindle, while the others read the same text but in the traditional paper-printed format. Afterwards, the students were quizzed on what they could remember about the story. The paper readers beat the e-readers in almost every measure (*see graph*), most significantly in recalling the order in which the narrative unfolded. So... it seems you take in more of the plot when you read it on paper.

The same seems to be true of reading textbooks. Naomi Baron – a professor of linguistics at the American University in Washington, DC – surveyed students in the US, Germany, Japan and Slovakia and asked which medium they find it easiest to concentrate on when studying. Despite being offered cell phones, laptops, tablets and e-readers, ninety-two per cent of respondents opted for a hard copy. It seems books offer the least amount of distraction and the fewest possible avenues for procrastination.

As many of us read in bed before going to sleep, scientists have also looked into the effect of e-readers on our slumber. Researchers from Harvard Medical School conducted a study, published in the journal *PNAS*, in which participants were asked to read in the four hours before bed over a period of two weeks. Those reading e-books were found to fall asleep ten minutes later than their paper-reading counterparts. The e-readers also experienced significantly less rapid eye movement (REM) sleep – the most restorative part of our sleep cycle. No wonder they also took longer to reach the same level of alertness the next morning compared to the book readers.

So, the upshot of the research into e-readers versus books seems to be this: e-readers can be a lot more convenient, but perhaps stick to them for shorter texts. If you're reading something more in depth that you want to fully comprehend, or you are reading in the evening, stick to the good old-fashioned printed page.

WAYS TO IMPROVE YOUR WI-FI SIGNAL

These days, Internet access ranks alongside food and water as one of life's essentials. Being cut off from the online world can often feel like losing a limb. Second to this feeling is crummy Internet access.

YouTube videos and streamed movies that buffer, downloaded email attachments that take half a day to materialize. Sure, things are better now than the days of dial up – when you couldn't be on the phone and the web at the same time, and pictures would load one painstaking line at time – but a slow Internet connection is still a pain in the neck.

Thankfully, there are ways to avoid this very modern melancholy. But first up you should realize that Wi-Fi isn't the way to fast Internet. A 2011 study of nearly three thousand broadband users in the US found that their wireless Internet speed was on average 29.7 per cent lower than if they were using a wired connection such as an Ethernet cable. But if tethering your laptop to a router seems just a little too twentieth century for your liking, then Jason Cole from Imperial College, London, is the man for you.

The physicist has used computer modelling to calculate the optimum position to place that magic little box that is your home router. The result probably won't come as much of a surprise – place the router in the middle of your home. That way it is more likely to reach everywhere. The radio waves that carry all the LOL cats to your computer repeat every twelve centimetres, meaning that their strength is decreased by passing through walls. So, if you want the ultimate signal, make sure you are in line of sight of the router. This can be tricky, considering that plug sockets are often tucked away in the corners of rooms, so consider using an extension cable to allow you to position the router away from the wall. If you really want to geek out and find the perfect spot, download Cole's Android app called *WiFi Solver FDTD* and upload the floor plan of your dwelling to pin down a personalized optimum router location.

Keeping your router off the ground is important, too. Many modern devices project their signal downward, meaning the rats in the basement might be getting a better signal strength than you. Likewise, don't place your router near other electrical equipment. It can interfere with the radio waves and downgrade your signal strength.

Finally, consider changing the channel over which your router broadcasts the Wi-Fi signal. Particularly in big cities, where people live crammed together in apartment blocks, the airwaves can be similarly jammed. If everyone in the block is transmitting on the same frequency, then it is likely you are drowning each other out. Changing channels on the router is a fairly painless task, with many online guides for how to do it. But given that there are often at least eleven channels to choose from, which one should you pick? The key here is to download some free software that can scan your local environment and show you which channel is carrying the least traffic – that's often the one you want.

HOW TO CHOOSE A GOOD PASSWORD

Despite years of advice urging us to beef up the strengths of our passwords, it seems not many of us are actually taking note. Web hosts WP Engine got their hands on a database of ten million leaked passwords that had been compiled by security consultant Mark Burnett. Their analysis showed that 0.6 per cent of the passwords were simply "123456"...

The most common passwords included "password" and "qwerty" and the ten most popular alone accounted for sixteen out of every one thousand passwords; 8.4 per cent ended with a number between zero and ninety-nine and more than twenty per cent of the time that number was "one".

WP Engine also found that we humans are suckers for patterns. Whilst "1qaz2wsx" may look like an impressively rigorous password, it becomes less so when you realize that it was created with adjacent keys. Creating passwords in this way is known as a "keyboard walk", and hackers running such combinations could crack it pretty easily. "Adgjmptw" featured in the top twenty keyboard walks, but is the only one that is not a walk across a QWERTY keyboard. Can you figure out what it is (answer at the end)?

However, if those are all examples of bad passwords, just what makes a good one? Especially

when you factor in the need to be able to remember it. In a 2011 study, Saranga Komanduri and colleagues at Carnegie Mellon University sought out the answer. Participants created a total of twelve thousand passwords based on a variety of construction rules, including "comprehensive8", in which passwords had to be at least eight characters long, contain upper and lower case letters, a number, a symbol and not contain a dictionary word. For example, "Tgfq1&Ha".

If you are thinking those rules are complicated, then you'd be right. The researchers found that only eighteen per cent of participants could create a suitable comprehensive8 password on their first attempt. In fact, twenty-five per cent of people gave up before they successfully created a working password. Of course, such efforts would be rewarded if they led to a greater level of security. So, Komanduri put comprehensive8 up against other password

8.4% OF PASSWORDS END WITH A NUMBER BETWEEN 0 AND 99.

THE 50 MOST USED PASSWORDS

1. 123456	11. 123123	21. MUSTANG	31. 777777	41. HARLEY
2. PASSWORD	12. BASEBALL	22. 666666	32. F*CKY*U	42. ZXCVBNM
3. 12345678	13. ABC123	23. QWERTYUIOP	33. QAZWSX	43. ASDFGH
4. QWERTY	14. FOOTBALL	24. 123321	34. JORDAN	44. BUSTER
5. 123456789	15. MONKEY	25. 1234567890	35. JENNIFER	45. ANDREW
6. 12345	16. LETMEIN	26. P*S*Y	36. 1230WE	46. BATMAN
7. 1234	17. SHADOW	27. SUPERMAN	37. 121212	47. SOCCER
8. 111111	18. MASTER	28. 270	38. KILLER	48. TIGGER
9. 1234567	19. 696969	29. 654321	39. TRUSTNO1	49. CHARLIE
10. DRAGON	20. MICHAEL	30. 1QAZZWSX	40. HUNTER	50. ROBERT

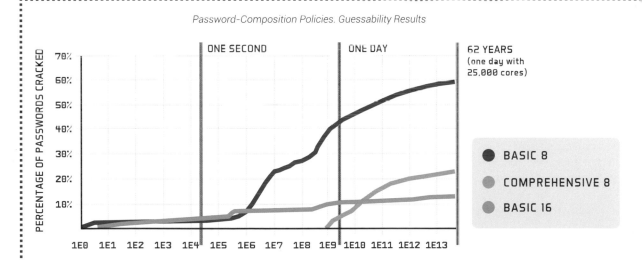

Password-Composition Policies. Guessability Results

ONE SECOND   ONE DAY   62 YEARS
(one day with 25,000 cores)

- ● BASIC 8
- ● COMPREHENSIVE 8
- ● BASIC 16

THE PASSWORD IS DYING

Our advice is all well and good, but we're afraid it does come with an expiry date. That's because it is very likely our children, and our children's children, will laugh relentlessly at us for the fact we ever had to type in some arbitrary string of letters and numbers in order to gain access to our most precious information.

We're already seeing the beginnings of such a revolution. In 2016, Facebook announced their Account Kit initiative at an industry conference. Rather than signing in using a password, you input your phone number instead. A confirmation code is then sent to your phone and that's what gets you in.

Smartphone banking apps are beginning to let you into your account by recognizing your fingerprint through a pad on your device. Major banks are also busy developing technology that goes a step further by authenticating your identity simply by the way you hold and use your phone. Face and iris recognition technology isn't that far over the horizon either.

It is no surprise, because people hate passwords. Surveys suggest that seven in ten people have to hit the "forgot my password" button twice a month. If our technology can know it is us without the inconvenience of having to provide explicit credentials, then we might one day look back on passwords with the same kind of nostalgia as the 8-bit characters so loving rendered in these pages.

creation methods, including basic8 and basic16 – passwords with a minimum of eight and sixteen characters respectively – but no other restrictions. The researchers then subjected these passwords to two different forms of hack.

The hardest to crack? Basic16. Even after ten billion guesses, these passwords were only hacked twelve per cent of the time. That compares to twenty-two per cent for comprehensive8 and sixty per cent for basic8. So, not only is the requirement for uppercase/lowercase, numbers and symbols more frustrating for the user, it seems it doesn't offer as much protection as a string of sixteen lowercase letters. So, Tgfq1&Ha isn't as a good as four random words strung together to make sixteen letters, for example redpiggolfcheese. Concocting a story around the words will help you remember it – a red pig hitting a lump of cheese with a golf club is a pretty hard image to shake!

ANSWER: Adgjmptw is a keyboard walk on a phone's number pad, created by pressing each of the numbers 2–9 in order.

QWERTYUIOP[]ASDFGHJKL:'\ZXCVBNM<>/
WERTYUIOP[]ASDFGHJKL:'\ZXCVBNM<>/Q
ERTYUIOP[]ASDFGHJKL:'\ZXCVBNM<>/QW
RTYUIOP[]ASDFGHJKL:'\ZXCVBNM<>/QWR

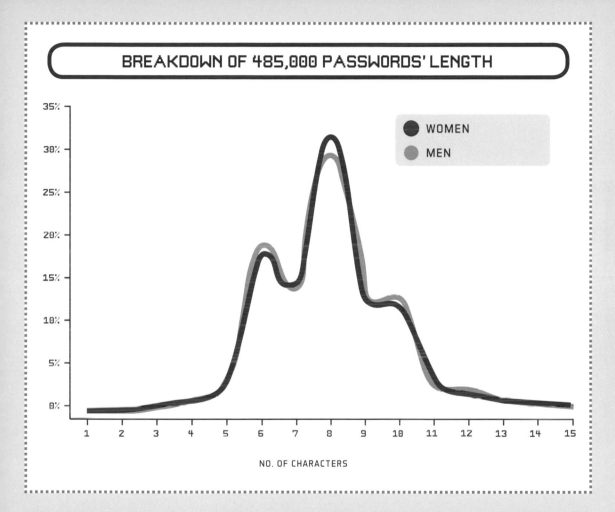

BREAKDOWN OF 485,000 PASSWORDS' LENGTH

WOMEN
MEN

NO. OF CHARACTERS

PASSWORDS CREATED USING **BASIC16** ARE THE HARDEST TO CRACK.

ARE VIDEO GAMES GOOD FOR YOU?

Video games are a staple of the geek kingdom, but the media often paint them in a bad light. We've all seen the headlines warning that the latest "shoot 'em up" is turning kids into a generation of violent, bloodthirsty thugs.

However, is that claim actually backed up by solid science? The honest answer is that the scientific community is divided. Some studies have shown a link between violent video games and an increased level of aggression. Others have found no link at all. However, it cannot be stressed highly enough that those finding an association could not say for certain that the aggression was caused by playing the games.

It is a typical "correlation versus causation" conundrum. Take the famous example of shark attacks and ice cream sales both increasing in Australia in the summer. It would be preposterous to say that buying more ice creams causes more shark attacks (or vice versa). Instead, there is a confounding variable at play – warmer weather is probably responsible for both increases. When it comes to video games, there may be some other variable involved, such as family background or mental health. A 2015 report by the American Psychological Association looked at one hundred studies on the subject and concluded that there was a link between violent video games and aggression, but that more work was still needed. The researchers went on to say that video games were just one of the risk factors associated with aggressive behaviour.

In fact, the amount of time spent playing video games, rather than their content, might be the thing to watch out for. According to a study by psychologists at the University of Oxford, and published in the journal *Psychology of Popular Media Culture*, kids spending more than three hours a day playing video games were more likely to be hyperactive, get involved in fights and be disinterested in school. They found the type of game did not matter. Conversely, they found that playing video games for up to an hour every day was beneficial.

This is part of a growing body of evidence that video games have many advantages. In 2013, the American Psychological Association reviewed research into the positive effects of gaming. Publishing their results in the journal *American Psychologist*, they found that shoot 'em ups improved a player's capacity to think about objects in three dimensions. A 2013 study also discovered that children playing strategic games were found to have improved both their problem-solving abilities and their school grades a year later. This idea was cemented in 2015, when a study published in Nature's *Scientific Reports* journal found that seasoned action gamers had increased grey matter and greater connectivity in their brains compared to non-gamers.

So, on balance, it seems that video games can be good for you, as long as you don't spend all your time playing them. It comes back to the old adage: everything in moderation. Which is good news for geeks everywhere!

TOP TIPS:

✓ Vary your gaming – strategy games may improve your problem-solving abilities, while action gamers have increased connectivity in their brains.

✗ Avoid spending too long gaming – it may have a negative effect on your wellbeing.

IS SOCIAL MEDIA GOOD FOR YOU?

How will future generations look back on today? What will the here and now be remembered for? The 1960s are most associated with free love, hippies and Woodstock. When you think of the 1970s, your mind probably conjures up images of disco and flares. Likewise, consider the 1980s, and shoulder pads, the New Romantics and shell suits maybe all pop up on the radar.

It is possible that the present will be remembered for the advent of widespread social media. Sure, Twitter and Facebook officially launched in the Naughties, but according to the Pew Research Center, only seven per cent of American adults were using social media back then. In 2015, it was sixty-five per cent By the end of that year, there were 1.5 billion Facebook users worldwide.

However, is this constant access to what everyone else is up to having a positive or a negative effect on us? In 2012, researchers at Salford University Business School in the UK surveyed nearly three hundred people about their social media habits. Twenty-seven per cent reported that using such sites had a negative effect on their behaviour. The reason given by many was that they felt less self-confident due to comparing themselves to others. The idea that our sense of self-worth is largely based on social comparison is nothing new – the idea dates back as far as the 1950s, to the work of psychologist Leon Festinger. But arguably it has never been so in our faces as in the age of social media.

What you have to keep in mind is that your friends' social media stream is unlikely to be a fair reflection of their lives – it is a cherry-picked highlight reel of what they choose to present to the world. It is well known that people are more likely to share positive emotions whilst hiding the negative ones. In 2011, a study published in the journal *Personality and Social Psychology Bulletin* not only backed this up, but found that we overestimate the positive aspects of other people's lives. It is a double whammy: not only are your social media friends presenting their best bits, but you are not very good at evaluating them.

So, comparing your life to a fleeting insight into the life of someone you haven't seen since your primary school days is not healthy. In fact, a 2014 study published in the *Journal of Social and Clinical Psychology* found that such comparisons were behind the link between more time spent on Facebook and depressive symptoms. A separate study by researchers at the University of Cologne, and published in the same journal, found that this effect was greater in those who are already prone to depression and low self-esteem.

The problem can be particularly acute for teenagers. A 2015 report by the UK's Office for National Statistics found increased levels of depression among adolescents who spent more time engaged with social media. They found that twelve per cent of children who spend little or no time on sites like Facebook and Twitter have mental health problems. However, this rises to twenty-seven per cent for those spending more than three hours a day on such activity. Of course, we should be careful leaping to the correlation/causation fallacy that it is definitely the social media causing the depression, but we have already seen that those who are depressed are not helped by the easy comparison to others that social media exposes us to. A separate 2015 study, this time by Heather Cleland Woods at the University of Glasgow, linked an overuse of social media with poorer sleep quality, lower self-esteem and higher levels of anxiety and depression.

What about the number of online friends you have? How do they compare to face-to-face relationships? Not well, it turns out. In 2013, researchers surveyed five thousand Canadians about topics including their overall wellbeing and their social media activities. Their first conclusion was that the number of real-life friends we have boosts our subjective wellbeing, even after controlling for income, variation in demographics and personality differences. For example, doubling the number of friends we have in the real world has the same effect on our happiness as a fifty per cent increase in earnings. However, the number of friends you have online is largely uncorrelated with happiness. The results, which were published in the journal *PLoS ONE*, found that the effect of real-life friends was particularly stark among single, divorced, separated or widowed people than for those married or co-habiting.

In the face of all this evidence, the case against social media looks almost iron-clad. And yet we shouldn't be too hasty to jump to conclusions. Back in 2012, child advocacy group Common Sense Media surveyed over one thousand thirteen- to seventeen-year-olds. Here is a selection of what they found:

SOCIAL MEDIA MAKES THEM...	PERCENTAGE
... FEEL LESS SHY	29%
... FEEL MORE SHY	3%
... MORE OUTGOING	28%
... LESS OUTGOING	5%
... MORE CONFIDENT	20%
... LESS CONFIDENT	4%
... BETTER ABOUT THEMSELVES	15%
... WORSE ABOUT THEMSELVES	4%

So, as far as teenagers themselves are concerned, the good seems to outweigh the bad. And it could be that all of us feel a little less lonely thanks to our social media posts. That is according to a 2013 study published in the journal *Social Psychological and Personality Science*. Researchers split participants into two groups whilst simultaneously monitoring the moods and feelings of both. One group was asked to post more status updates to Facebook than they normally would, and the other group weren't given any instructions at all. The first group were found to have experienced a greater decrease in their feelings of loneliness by the end of the study period.

Social media is a relatively new addition to modern life, and researchers are still getting to grips with its impact. Whilst the jury remains out on whether it is good or bad for you, it seems clear that you should take what you see in your friends' timelines with a pinch of salt and spend more time cultivating offline friendships than online ones.

HOW TO BE MORE SUCCESSFUL ON TWITTER

Along with Facebook, Twitter has changed the way we interact online. Whilst Facebook was initially just about maintaining contact with people we knew, Twitter opened up access to practically anyone. It is little surprise, then, that celebrities are the most followed people on the site, with the most successful having tens of millions of followers.

Now the chances are you are not a celebrity, and you probably don't have many followers either — eighty per cent of Twitter users have fewer than ten followers. According to a 2013 report by O'Reilly Media, having one thousand followers puts you in the top four per cent of tweeters. So how can you climb this social media ladder?

In 2013, computer scientist C. J. Hutto, of the Georgia Institute of Technology, led a study analysing over half a million tweets and looking at aspects of Twitter behaviour which lead to an increase or decrease in followers.

First, here are the factors that they found will decrease your followers:

- **Only broadcasting, not discussing. The power of Twitter is in its ability to be a tool for discussion. Not mentioning other people, or not replying to their tweets, saw an exodus of hard-won followers.**
- **Being overly negative. People want to hear good news more than they do bad.**
- **Hashtag abuse. A well-chosen hashtag can work wonders. But, like, #totally hashtagging the #hell out of your #tweet is annoying #awkward #twitteradvice #lol.**
- **Talking only about yourself. Tweets with lots of "I" and "me" in them saw a drop in followers.**

If you are already avoiding these pitfalls, then you are off to a good start. But according to Hutto's study, there are positive things you can do to grow your followers:

- **Get retweeted. If your existing followers are sharing your content among their network, then you are increasing your exposure to potential new followers (more on this later).**
- **Become a source of information. That means sharing links to interesting content elsewhere on the web. The boost associated with posting information was found to be thirty times the detrimental effect of using "I" and "me" all the time.**
- **Make it personal. Users whose profile had a detailed biography and a link to a website had more followers.**

Do all of these things, and you will be well on your way to Twitter success. But just how do you get people to retweet (RT) your content? That's the question Bongwon Suh and colleagues at the Palo Alto Research Center in California set out to answer. They analysed seventy-four million tweets and what they found backs up much of the advice we have already given. A key factor was the inclusion of a URL or hashtag (not too many of course!). People are far more likely to share a link to an interesting article than your description of what you had for lunch. The other thing you could try is simply asking people to retweet your content – a tactic that is surprisingly effective. According to self-styled social media scientist Dan Zarrella, include "Please Re-Tweet" and it is nearly five times as likely to happen.

Of course, if you want to expand your Twitter network, you could always follow @geekguidetolife, along with your authors @skyponderer and @munkeatlooi.

CHECK OUT WWW.GEEKGUIDETOLIFE.COM FOR SOME AWESOME ADVICE ✓

HERE'S TO BEING MORE SUCCESSFUL ON TWITTER #GEEKGUIDETOLIFE ✓

THANKS @GEEKGUIDETOLIFE FOR YOUR ADVICE! ✓

SO #YOU #THINK #YOU #CAN #TWEET? #OMG #NOTCOOL #HASHTAG ✗

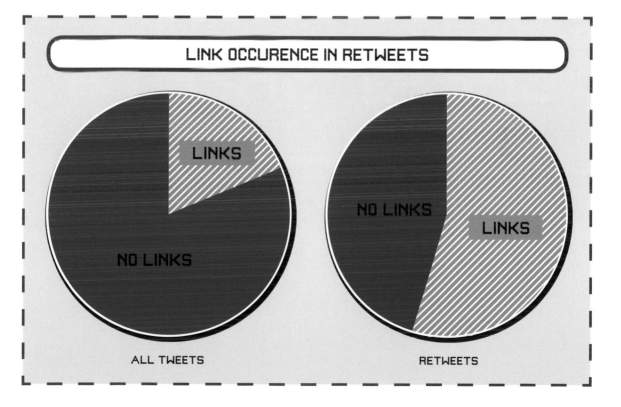

LINK OCCURENCE IN RETWEETS

LINKS

NO LINKS

ALL TWEETS

NO LINKS

LINKS

RETWEETS

CUT-OUT-
AND-KEEP
CHEAT SHEETS

WHAT'S THE BEST WAY TO GET KETCHUP OUT OF THE BOTTLE?

- Turn the bottle upside down with the lid on.
- Shake a couple of times. Remove the lid and pour.

HOW TO ROCK AT ROCK, PAPER, SCISSSORS

- Keep your moves random.
- Play with your eyes closed.

HOW TO BOIL THE PERFECT EGG

- Place in a water bath at a temperature between 60 and 66°C.
- Consult graph for length of cooking.

HOW MUCH SEX SHOULD YOU BE HAVING?

- Around once a week improves mood.
- Level of happiness does not increase with any more than that.

HOW TO COOL DOWN ON A HOT DAY

- Hot drinks can cool you down more effectively than cold ones – if you're stationary.
- Avoid drinks like coffee that dehydrate you.

SHOULD YOU TRUST THE "FIVE-SECOND RULE"?

- Not if you're a child or elderly.
- Not if you're picking up moist food from a tiled floor.

HOW TO KICK SOME SERIOUS ASS AT MONOPOLY

- Buy orange properties.
- Snap up the stations.

HOW TO BE DEADLY AT DARTS

- Throw at random if you regularly score less than 39 with three darts.
- Aiming for 7 is better than 20 for reasonably accurate players.

HOW TO MAKE THE PERFECT CUP OF TEA

- Loose-leaf tea brewed in a pre-warmed teapot for three minutes.
- Milk in mug first.

HOW TO SPOT A LIAR

- Distracted judgements can increase the accuracy.
- Be wary of unconscious bias.

HOW TO EAT LESS

- Imagine eating lots of the food you are about to eat.
- Avoid snacks during emotionally distressing films.

WHERE SHOULD YOU SIT AT THE MOVIES?

- Two-thirds of the way back and in the middle.
- If unavailable, left-hand side less likely to be crowded.

WANT TO WIN AT SPORTS? WEAR RED

- Does what it says on the tin.

HOW TO BECOME A MASTER OF STONE SKIMMING

- Launch a flat, round, spinning stone at twenty degrees to the water at at least 2.5 metres per second.
- The first bounce should occur within 4.5 metres.

THE BEST TIME TO SETTLE DOWN

- Date and reject the first thirty-seven per cent of your likely good matches.
- Adjust to just thirty per cent if you're ok with "good" rather than "great".

HOW TO QUIT BITING YOUR NAILS

- Distract yourself: chew gum, play with a ball.
- Physically stop yourself – sit on your hands.

HOW TO STOP A TUNE PLAYING OVER AND OVER IN YOUR HEAD

- Reactivate your mind by doing a Sudoku puzzle.
- Chew gum.

HOW TO NAVIGATE WITHOUT GPS

- Above the equator find north using the star Polaris. Below the equator use the Southern Cross.
- During the day use a shadow stick.

WHAT SCIENCE KNOWS ABOUT SUCCESSFUL MARRIAGES

- Say "thank you".
- Use couple-identifying words like "we".

HOW TO STOP YOUR EARPHONES TANGLING

- Clip the wires together just below the earbuds.

WHAT'S BETTER: BOOK OR E-READER?

- People comprehend more when reading a physical book.
- E-readers can prevent you from sleeping.

HOW TO DEAL WITH A BREAK-UP

- Don't wallow, but take some time to reflect on the experience.
- Eat chocolate.

HOW TO ARRANGE THE FOOD IN YOUR FRIDGE

- Raw meat and fish on the bottom shelf.
- Don't put milk in the door or fruit and veg on top of the fridge.

HOW BEST TO STACK THE DISHWASHER

- Carbohydrate-stained items in centre of top rack. Protein-stained items in a circular configuration on bottom shelf.
- Don't pre-rinse, overfill or block rotating arm.

HOW TO CURE A HANGOVER

- Don't mix drinks.
- Make sure you eat, and drink plenty of water.

HOW TO BUILD THE BEST SANDCASTLE

- Don't build with bucket and spade – compact sand with your hands.
- Triple the base width to double the height.

HOW TO SUCCEED ON TINDER (AND OTHER DATING SITES)

- Concentrate on your picture.
- Write seventy per cent about you, thirty per cent about what you're looking for.

HOW TO BUILD THE PERFECT PAPER PLANE

- Use A4 100gsm laser paper.
- Make crisp folds and follow the 165 degree/155 degree/165 degree angle rule.

HOW TO STOP CRYING OVER YOUR ONIONS

- Cut a chilled onion with a sharp knife and don't cut into the root.
- Consider working with a fan.

WHICH COLOUR CAR SHOULD YOU BUY?

- Anything but black.

HOW TO MAKE FRIENDS THROUGH KARAOKE AND DANCING

- Sing or dance to increase camaraderie within a group.
- On the dancefloor, move your head and torso in time to the groove.

WHAT'S THE BEST WAY TO COMMUTE TO WORK?

- Don't drive unless you have to.
- Walkers are happiest.

THE BEST STRATEGIES FOR BUYING AND SELLING ON EBAY

- Buyers should use sniping software and not bid on listings with a red background. They should also negotiate on items with round starting prices.
- Sellers should do the opposite.

SHOULD YOU WALK OR RUN IN THE RAIN?

- Run as fast as is safe.
- Unless the rain is coming from behind you, in which case match the speed of the wind.

HOW TO BE A BETTER HAGGLER

- Sit in a hard seat and make sure your opening offer isn't a round figure.

HOW TO MAKE FOOD TASTE BETTER

- Play the appropriate music.
- Use heavy cutlery and serve on a round, white plate.

HOW TO BE A BETTER SAVER

- Think of time in a cyclical way rather than a linear one.
- Set savings targets in terms of days not years.

WHY YOU SHOULD GIVE MONEY TO CHARITY

- It gives you the same buzz as eating chocolate or having sex.

HOW TO MANAGE JET LAG

- Stagger your sleep and light exposure in the week before your flight.
- Avoid caffeine and alcohol.

HOW TO SPEND LESS IN RESTAURANTS

- Be aware of the power of anchor items on menus.
- Be wary of gifts from blonde waiters.

SPEND MONEY ON EXPERIENCES, NOT THINGS

- You'll be happier.

HOW MUCH MONEY DO YOU NEED TO BE HAPPY?

- Any income above around $75,000 doesn't boost happiness further.
- Be the richest person in a poor area, not the poorest person in a rich one.

THE GEEK GUIDE TO COLDS AND FEVERS

- Drink hot drinks, especially soups.
- If you have to blow your nose, blow gently, and one nostril at a time.

WAYS TO IMPROVE YOUR WI-FI SIGNAL

- Place router in the centre of the house, away from walls and electrical equipment.
- Change router frequency to less crowded channel.

HOW TO SPLIT THE RESTAURANT BILL

- Everyone should pay for what they've eaten.
- Alternatively use Matt Parker's "Standard Meal Units".

HOW TO BE A BETTER PUBLIC SPEAKER

- Visualize the steps necessary for a successful talk.
- Have penetrative sex in the days/weeks running up to the talk.

HOW TO BE MORE CONFIDENT

- Stand in a superhero power pose whilst listening to Queen's "We Will Rock You".

HOW TO IMPROVE YOUR MEMORY

- Remember someone's name by saying it aloud to another person.
- Chew gum, drink coffee and doodle.

HOW TO CHOOSE A GOOD PASSWORD
- Avoid "123456", "password" or keyboard walks.
- Sixteen letters is better than a mixture of eight letters and numbers.

IS SOCIAL MEDIA GOOD FOR YOU?
- Don't compare your to life to a fleeting glimpse of someone else's.
- Offline friendships often contribute more to our wellbeing than online ones.

ARE VIDEO GAMES GOOD FOR YOU?
- No concrete link between action video games and increased violent behaviour.
- Limit your time playing games rather than the content of the game itself.

HOW TO BE MORE SUCCESSFUL ON TWITTER
- Have a conversation – don't just talk about yourself.
- Include links, a limited number of hashtags and ask to be retweeted.

HOW TO SCIENCE YOUR WAY TO SUCCESS IN A JOB INTERVIEW
- Firm opening handshake with a warm hand.
- Consider wearing black or blue, but avoid orange.

HOW TO NETWORK YOUR WAY TO CAREER SUCCESS
- Be in a large, open network, not a small, closed one.
- The most successful people act as "brokers" between networked groups.

HOW TO STOP PROCRASTINATING
- Don't beat yourself up for past procrastination.
- Avoid "analysis paralysis" by throwing yourself in at the deep end and just getting started.

SHOULD YOU CHANGE LANES IN A TRAFFIC JAM?
- Yes, as long as the traffic density isn't super high.
- Switching lanes often increases risk of accident, however.

WHY YOU SHOULD AVOID BEING "HANGRY"
- If you're in a bad mood, check when you last ate.
- Try reaching for a small snack when irritable.

HOW TO SAVE MONEY IN THE SUPERMARKET
- Make a list and stick to it.
- Play your own music, use a small trolley and look at all levels of shelving.

HOW TO LEARN ANOTHER LANGUAGE
- Listen to new words repetitively and get feedback on your pronunciation.
- Immerse yourself with native speakers.

HOW TO BE MORE PERSUASIVE
- Give someone a reason to do something – say "because" – and ask don't order.
- Get them nodding along with you.

WHEN TO BUY AN AIRLINE TICKET (AND WHERE TO SIT ON BOARD)
- Buy around seven weeks before and not on a Friday.
- Sit in the back third of the plane, on the aisle and within five rows of an exit.

HOW MUCH SLEEP YOU REALLY NEED
- Stick to a regular schedule of sleep and waking, even at weekends.
- Coffee nap: down a caffeinated beverage of your choice, then take a twenty-minute snooze.

HOW MUCH EXERCISE YOU REALLY NEED
- Do something physically active for fifteen minutes a day.
- Pace yourself.

WHEN TO TEXT NEXT
- ALWAYS wait.
- Around ten to sixty minutes should do it.

INDEX

CREDITS AND SOURCES

PICTURE CREDITS

Original illustrations by Phil O'Farrell
www.philofarrell.co.uk

Illustrations on pages 13 (top right) and 29 (left) courtesy of Rebecca Wright.

All other images courtesy of Shutterstock.com

DATA SOURCES

13 (top left): Thermal Ergonomics Laboratory, School of Human Kinetics, University of Ottawa, Ottawa, Ontario, Canada.

29: US National Sleep Foundation

30: Brant P. Hasler, Thomas W. Kamarck, Stephen B. Manuck, Matthew F. Muldoon and Patricia M. Wong

31: School of Psychology, Flinders University, Adelaide, Australia.

35: Annual Population Survey, ONS

43: Noam Sobel, Weizmann Institute of Science, Israel

44: Dan Ariely and Klaus Wertenbroch, "Procrastination, Deadlines, and Performance: Self-Control by Precommitment", *Psychological Science*, 2002

45: Michael J. A. Wohl, Timothy A. Pychyl, Shannon H. Bennett, *Personality and Individual Differences*

48: Ronald Burt, University of Chicago Booth School of Business

50: Albarracín, Dolores, Noguchi, Kenji and Senay, Ibrahim, *Psychological Science*

52: Cristian Danescu-Niculescu-Mizil, Lillian Lee, Vlad Niculae and Chenhao Tan

56: George Loewenstein et al, "Does Increased Sexual Frequency Enhance Happiness?", *Journal of Ecomomic Behavior & Organization*, 2015

75: "Couples, the Internet, and Social Media", Pew Research Center

82: Ruben Mercadé-Prieto and César Vega, "Culinary Biophysics: on the Nature of the 6XC Fgg", *Food Biophysics*, 2011

97: Alberto Cruz and Barry G. Green, "Thermal stimulation of taste", *Nature*, 2000

117: Russell Hill and Robert Barton, University of Durham, 2005

119: Lydéric Bocquet, *American Journal of Physics*

127: Monash University, Australia

138 (top): *Expedia 2015 Air Travel Trends*

138 (bottom): Ian Savage, Northwestern University

140: Lobkovsky Meitiv, *Playing With Models*

143: Daniel Bonn, Mehdi Habibi, Peder Møller and Maryam Pakpour, *Scientific Reports*, 2012

149: Nancy Puccinelli, Rajesh Chandrashekaran, Dhruv Grewal and Rajneesh Suri (2013) "Are Men Seduced by Red? The Effect of Red Versus Black Prices on Price Perceptions", *Journal of Retailing*, 89 (2) pp. 115–125.

151: *Journal of Experimental Social Psychology*/Malia Mason

152: *Journal of Marketing Research*/DILIP SOMAN and MIN ZHAO

157: *Environment & Behaviour*/Adrian North and Amber Shilcock

158: *Psychological and Cognitive Sciences*/Angus Deaton and Daniel Kahneman

161: *Science*/Christian Elger

163: *Journal of Personality and Social Psychology*/ Leaf Van Boven and Thomas Gilovich

165: *PLoS ONE*/Elizabeth Dunn et al

169: Matthew Backus, Thomas Blake and Steven Tadelis

171: Robert Matthews/*School Science Review*

172: Anne Mangen and Jean-Luc Velay

177 (top): WP Engine

177 (bottom): Saranga Komanduri et al

179: WP Engine

183: Common Sense Media

185: Bongwon Suh et al